Density Functional Theory in Quantum Chemistry

Takao Tsuneda

Density Functional Theory in Quantum Chemistry

Springer

Takao Tsuneda
University of Yamanashi
Kofu, Japan

ISBN 978-4-431-54824-9 ISBN 978-4-431-54825-6 (eBook)
DOI 10.1007/978-4-431-54825-6
Springer Tokyo Heidelberg New York Dordrecht London

Library of Congress Control Number: 2014930332

Printed on acid-free paper

Springer is part of Springer Science+Business Media (www.springer.com)

Preface

Density functional theory (DFT) was developed to calculate the electronic states of solids containing huge numbers of electrons. In the earliest years, DFT was, therefore, used only for calculations of band structure and other properties of solids. However, DFT began to be used in quantum chemistry calculations in the 1990s, and today it has become the predominant method, accounting for more than 80 % of all quantum chemistry calculations, after only two decades. Quantum chemistry is aimed mainly at chemical reactions and properties. Because chemical reactions are usually associated with electron transfers between much different electronic states, highly sophisticated methods are required, incorporating high-level electron correlations of well-balanced dynamical and nondynamical correlations (see Sect. 3.2) to quantitatively reproduce the reactions. Quantum chemists have, therefore, focused on how to incorporate high-level electron correlations efficiently for several decades. So far, various methods have been developed with the difference mainly in the approaches for sorting out electron configurations to incorporate electron correlations efficiently. Prior to DFT, conventional methods have required much computational time, making it difficult to calculate the electronic states of large molecules, even for those containing several dozen atoms in the 1990s. The appearance of DFT altered this situation. Because DFT incorporates high-level electron correlations of well-balanced dynamical and nondynamical correlations simply in exchange-correlation functionals of electron density (see Sect. 4.5), it enables us to calculate chemical reaction energy diagrams quantitatively, with computational times equivalent or less than those for the Hartree–Fock method.

In this book, the fundamentals of DFT are reviewed from the point of view of quantum chemistry. The fundamentals of DFT have so far been described in many reference books. However, most DFT books explain the fundamentals of conventional DFT methods used in solid state calculations, which are not necessarily the same as those used in quantum chemistry calculations. In order to figure out how to use DFT to approach quantum chemistry, it is necessary to know the meaning of electron correlation and the strategies to incorporate high-level electron correlations. Molecular orbital energy is one of the most reliable indicators to test the balance of the electron correlations, which are mostly included in exchange-correlation functionals. Based on this concept, this book first introduces the history and fundamentals of quantum chemistry calculations, then explains exchange-correlation functionals and their corrections especially for incorporating high-level electron correlations, and finally describes highly sophisticated DFT methods to provide correct orbital energies.

The objectives and outlines of each chapter are as follows:

In Chap. 1, DFT is placed in the history of quantum chemistry, and then the Schrödinger equation and the quantizations of molecular motions are reviewed. First, the history of quantum chemistry is overviewed to place DFT in the history of quantum chemistry. This chapter then reviews the backgrounds and fundamentals of the Schrödinger equation with the meaning of the wavefunction, in accord with the history. As the first applications in quantum chemistry, the quantizations of the three fundamental molecular motions are discussed using simple models, especially for the meanings of the Schrödinger equation solutions.

According to the history of quantum chemistry, the Hartree–Fock method and its computational algorithms are introduced in Chap. 2. First, the Hartree method and molecular orbital theory are briefly reviewed as the foundations of molecular electronic state theories. Based on these, the Slater determinant for the wavefunction and the Hartree–Fock method based thereon are then explained. As the computational algorithms of the Hartree–Fock method in quantum chemistry calculations, this chapter also describes the Roothaan method, basis functions centering on Gaussian-type functions, and high-speed computation algorithms of the Coulomb and exchange integrals. The unrestricted Hartree–Fock method for open-shell system calculations is also surveyed. This chapter also explains the electronic configurations of the elements in the periodic table, confirmed by the Hartree–Fock method to a considerable extent.

Chapter 3 reviews electron correlation, to which the highest importance has been attached in quantum chemistry, for the meaning and previous approaches to incorporate it. After describing the main cause for electron correlation, dynamical and nondynamical electron correlations are introduced to clarify the details of electron correlation. As the calculation methods for these electron correlations, this chapter briefly reviews the configuration interaction and perturbation methods for dynamical correlations and the multiconfigurational self-consistent field (SCF) method for nondynamical correlations. This chapter also mentions advanced electron correlation calculation methods to incorporate high-level electron correlations.

In Chap. 4, the Kohn–Sham equation, which is the fundamental equation of DFT, and the Kohn–Sham method using this equation are described for the basic formalisms and application methods. This chapter first introduces the Thomas–Fermi method, which is conceptually the first DFT method. Then, the Hohenberg–Kohn theorem, which is the fundamental theorem of the Kohn–Sham method, is clarified in terms of its basics, problems, and solutions, including the constrained-search method. The Kohn–Sham method and its expansion to more general cases are explained on the basis of this theorem. This chapter also reviews the constrained-search-based method of exchange-correlation potentials from electron densities and the expansions of the Kohn–Sham method to time-dependent and response property calculations.

Exchange-correlation functionals, which determine the reliability of Kohn–Sham calculations, are compared in terms of the basic concepts in their development, and for their features and problems, in Chap. 5. This chapter uses as examples the major local density approximation (LDA) and generalized gradient approximation (GGA) exchange-correlation functionals and meta-GGA, hybrid GGA, and semi-empirical functionals to enhance the degree of approximation in terms of their concepts, applicabilities, and problems.

Chapter 6 reviews physically meaningful corrections for the exchange-correlation functionals, including their formulations and applications. As the specific types of corrections, this chapter covers long-range corrections, enabling us to calculate orbital energies and exchange integral kernels correctly; self-interaction corrections, improving the descriptions of core electronic states; van der Waals corrections, which are required in calculating van der Waals interactions; relativistic corrections, which are needed in the electronic state calculations of heavy atomic systems; and vector-potential corrections, which play a significant role in magnetic calculations.

Chapter 7 focuses on orbital energy, which is the solution of the Kohn–Sham equation and one of the best indicators to evaluate incorporated electron correlations, including the various approaches to reproduce accurate orbital energies. The physical meaning of orbital energy is first explained on the basis of the Koopmans and Janak theorems. Then, this chapter summarizes previous discussions on the causes of poor-quality orbital energies given in Kohn–Sham calculations and shows highly sophisticated exchange-correlation potentials, which have been developed to calculate accurate orbital energies. Finally, the long-range corrected Kohn–Sham method, which reproduces accurate occupied and unoccupied orbital energies simultaneously, is discussed, revealing the path to obtain accurate orbital energies.

This book has as its target readership the following groups: graduate students who are beginning their study of quantum chemistry, experimental researchers who intend to study DFT calculations from the beginning, theoretical researchers from different fields who become attracted to DFT studies in quantum chemistry, and quantum chemists who wish to brush up their fundamentals of quantum chemistry and DFT or wish to have a reference book for their lectures. Therefore, this book was designed to be useful in studying the fundamentals, not only of DFT but of quantum chemistry itself. Unlike representative DFT books such as Parr and Yang's *Density-Functional Theory of Atoms and Molecules* (Oxford Press) and Dreizler and Gross's *Density Functional Theory: An Approach to the Quantum Many-Body Problem* (Springer), this book explains DFT in practical quantum chemistry calculations using the terminology of chemistry. Because this book focuses on quantum chemistry, it basically omits DFT topics unrelated directly to quantum chemistry calculations. The detailed derivations of formulations are also neglected in this book, unlike many DFT books in physics, because this book is intended to instill the comprehension of DFT fundamentals. For the details required in the development of specific theories and computational programs, the reader is directed to the relevant papers that are cited.

Finally, I would like to acknowledge Prof. Donald A. Tryk (University of Yamanashi) for supervising the English translation and for giving productive advice. I would also like to acknowledge Prof. Andreas Savin (Université Pierre et Marie Curie and CNRS) for giving fruitful comments and discussions of Chap. 7. This book is basically the translation of my Japanese book, the English title of which is *Fundamentals of Density Functional Theory* (Kodansha). Again, I would like to record my thanks to Prof. Haruyuki Nakano (Kyushu University), Prof. Tetsuya Taketsugu (Hokkaido University), Prof. Shusuke Yamanaka (Osaka University), and Prof. Yasuteru Shigeta (Osaka University) for their detailed reviews of the Japanese version. Finally, I would like to express my thanks to Taeko Sato and Shinichi Koizumi (Springer, Japan) for providing the opportunity to publish my book and for waiting for the completion of my manuscript.

Kofu, Japan
November 2013

Takao Tsuneda

Contents

Chapter 1
Quantum Chemistry

1.1 History of Quantum Chemistry

Quantum chemistry is a branch of chemistry in which chemical phenomena are elucidated deductively on the basis of quantum mechanics. Chemistry covers a wide range of scales, from atoms and small molecules to large systems such as biomolecules and solids, and includes their structures, properties, and reactions. Except for statistical–mechanical factors, quantum mechanics controls chemistry. Actually, the electronic structure of matter is determined by solving the Schrödinger equation precisely, except for the contributions of relativistic effects. This seems to indicate that quantum chemistry is a subset of quantum mechanics designed for the study of chemistry. However, it should not be overlooked that *quantum chemistry focuses on the applications of chemistry and therefore traces the progress of experimental chemistry.* In fact, quantum chemistry has been synchronized with the progress of experimental chemistry. To make this clear, I briefly summarize the history of quantum chemistry with a focus on density functional theory (DFT) for electronic structure calculations below. Since this history intends to review the broad flow of progress in quantum chemistry, I would like to emphasize beforehand that it is neither objective nor exhaustive.

After the development of the Schrödinger equation, the main subjects of quantum chemistry have progressed roughly in six stages (see Table 1.1). Let us review the progress of quantum chemistry with consideration of specific major topics in experimental chemistry for each research stage.

First Stage: Fundamental Theories of Quantum Chemistry (1926–1937)

After the development of *the Schrödinger equation* (Schrödinger 1926) various significant fundamental theories of quantum mechanics were produced in a remarkably short period. Through the uncertainty principle (Heisenberg 1927) and Bohr's wave-

T. Tsuneda, *Density Functional Theory in Quantum Chemistry*,
DOI 10.1007/978-4-431-54825-6_1,

Table 1.1 Research stages of quantum chemistry and corresponding main subjects

Stage	Period	Main subject
1st	1926–1937	Fundamental theories of quantum chemistry
2nd	1950–1960	Customized theories available on computers
3rd	1961–1969	Approximate theories for calculating specific systems
4th	1970–1984	Quantum chemistry calculation programs and DFT
5th	1985–1995	Potential functionals and excited state theories
6th	1996–?	Easy-to-use theories focusing on utility

particle complementarity principle (lecture in Como, Italy, 1927), *the relativistic Schrödinger equation (Dirac equation)* (Dirac 1928) was developed 2 years later. A bunch of experiments were then carried out to support these fundamental theories, forming the concepts of quantum mechanics. Fundamental theories of quantum chemistry were also rapidly developed in this period as approaches for clarifying chemistry based on quantum mechanics.

The first target of quantum chemistry was *how to solve the Schrödinger equation for electronic motions in molecules*. To address this challenge, *the Hartree–Fock method* (Hartree 1928) and its *variational method* (Slater 1928), *molecular orbital theory* (Hund 1926; Mulliken 1927), and the *Slater determinant* (Slater 1929) were developed, resulting in the *Hartree–Fock method* (Fock 1930; Slater 1930), which is accepted as the precursor of quantum chemistry. Soon afterward, *the configuration interaction (CI) method* (Condon 1930), *Møller–Plesset perturbation method* (Møller and Plesset 1934), and *multiconfigurational SCF method* (Frenkel 1934) were proposed to incorporate an effect neglected in the Hartree–Fock method, which is "electron correlation." The *time-dependent response Hartree–Fock method* (Dirac 1930) for response property calculations and the empirical *Hückel method* (Hückel 1930) for electronic structure calculations of organic molecules were also suggested around the same period. In the field of solid state physics, the *Thomas–Fermi method* (Thomas 1927; Fermi 1928), which subsequently became the fundamental concept of *DFT*, was developed to solve the Schrödinger equation for solid state systems containing enormous numbers of electrons. The first *local density approximation (LDA)* was proposed for kinetic energy in this method, and this led to the development of the first *LDA exchange functional* (Dirac 1930) and the first *generalized gradient approximation (GGA)* for kinetic energy (von Weizsäcker 1935).

After a few test calculations of electronic structures, quantum chemistry immediately targeted *ways to interpret electronic state wavefunctions in molecules*. This is exemplified by the *hybrid orbital model* (Pauling 1928) for clarifying atomic orbitals in molecules, *Koopmans theorem* (Koopmans 1934) for giving orbitals meaning, *transition state theory* (Eyring 1935) for discussing chemical reactivity, the *linear combination of atomic orbital–molecular orbitals (LCAO–MO) approximation* (Lennard-Jones 1929; Coulson 1938), for modeling molecular orbitals, and the *chemical reaction principle* (Bell 1936; Evans and Polanyi 1938) for interpreting chemical reactions. Based on all of these components, fundamental theories of quantum chemistry were constructed to investigate the electronic structure of molecules.

Subsequently, the focus of many of the founders of quantum mechanics shifted to the atomic nucleus. However, following the discovery of nuclear fission in uranium in 1938 and the World War II from 1939 to 1945, the finest minds in physics were diverted to the design of atomic weapons. The war drastically decreased researchers' positions and funds, other than military-related projects, causing science to drift through a long period of stagnation. Quantum chemistry also hardly produced any remarkable studies except in atomic bomb-related studies until notable changes started to occur in the 1950s.

Second Stage: Customized Theories Available on Computers (1950–1960)

In the 1950s, *the appearance of computers* brought about a revolution in science and technology. The hardware development of computers made progress during this period: the world's first commercial computer, UNIVAC, was built in 1950, and the first large-scale scientific computer, the IBM701, appeared in 1952. Following the appearance of computers, researchers in many scientific fields started to seek new ways of applying computers to their fields. Quantum chemistry has been one of the scientific fields most affected by these developments.

Before the appearance of computers, it was usually not realistic to solve the Schrödinger equation for the electronic structures of molecules due to the huge numbers of computations. Since computers were able to bring the latter into reality, they triggered the rapid development of computational theories optimized for computer architectures. In particular, various theories and algorithms were proposed to make matrix operations available, because von Neumann-type computers, which are in the mainstream even at present (2013), are better suited for the matrix operations. Beginning with the development of *basis functions* (Boys 1950) for the matrix operations in 1950, the *Roothaan method* (Roothaan 1951; Hall 1951) was developed as a Hartree–Fock method utilizing basis functions, the *unrestricted Hartree–Fock (UHF) method* (Pople and Nesbet 1954) was proposed to extend the Roothaan method to open-shell electronic structure calculations, and the first *semiempirical calculation method* (Pariser and Parr 1953; Pople 1953) was suggested to approximate the Roothaan method with semiempirical parameters to speed up the calculations. The concept of *electron correlation* (Löwdin 1955) was suggested in this context. Various major analysis approaches were also developed, e.g., the *population analysis method* (Mulliken 1955) for investigating molecular orbitals calculated with basis functions and the *molecular orbital localization method* (Foster and Boys 1960) for making molecular orbitals close to hybrid orbital pictures. Furthermore, the *molecular dynamics method* (Alder and Wainwright 1959) was proposed in this stage to simulate structural transformations by approximating interatomic interactions with force fields.

Third Stage: Approximate Theories for Calculating Specific Systems (1961–1969)

In the 1960s, performance advances in precision measurement devices induced discoveries in many scientific fields, leading to the reconstruction of both theories and technologies.

In quantum chemistry, various theories were developed to calculate specific systems for making comparisons with precise experimental results. It was, however, difficult for computers at that time to obtain sufficiently accurate results for interesting molecules. Therefore, theories were usually derived with bold approximations available only for specific systems. The *extended Hückel method* (Hoffmann 1963) was first proposed by modifying the Hückel method, suggested 30 years earlier, in order to be applicable to chemical property calculations. Then, the orbital-based reaction analysis method of *frontier orbital theory* (Fukui et al. 1952) was applied to the extended Hückel method and succeeded in helping to reveal the mechanisms of Diels–Alder reactions in 1964. This was validated a year later by the *Woodward–Hoffmann rules* (Woodward and Hoffmann 1965). In this period, *Marcus theory* (Marcus 1956) was developed to explain electron transfer mechanisms in outer-sphere-type redox reactions. It is particularly worth noting in this period that by reconsidering the Thomas–Fermi method suggested 30 years earlier, similarly to the Hückel method, the *Hohenberg–Kohn theorem* (Hohenberg and Kohn 1964), which is the fundamental theorem of DFT, was proposed, and the *Kohn–Sham method* (Kohn and Sham 1965) was then suggested on the basis of this theorem in the field of solid state physics. That is, DFT was initially generated as a theory for calculating a specific system, i.e., that of the solid state.

As another trend in this period, it should be mentioned that major theories for highly accurate calculations of small molecules were developed following the increase of Gaussian basis functions for calculating polyatomic systems and dynamical and nondynamical electron correlation analysis (Sinanoğlu 1964) for clarifying the meaning of electron correlation. For example, the *cluster expansion theory* (Čížek 1966) and the *equation-of-motion (EOM) method* (Rowe 1968) were proposed. The multiconfigurational SCF method became popular in this period and led to the *multireference configuration interaction (MRCI) method* (Whitten and Hackmeyer 1969), which uses the MCSCF wavefunction as the reference function.

Fourth Stage: Quantum Chemistry Calculation Programs and DFT (1970–1984)

Computers began to make obvious contributions to the world's science and technology in the 1970s due to their widespread prevalence, as symbolized by the appearance of the first personal computer, the MITS Altair8800 (1974).

In quantum chemistry, various quantum chemistry calculation programs, including the *Gaussian* (1970) and *GAMESS* (1982) programs, which later became the major commercial and free programs, were released. A number of different frequently used algorithms were consequently developed, especially for speeding up calculations, during this period. For instance, the *Davidson matrix diagonalization algorithm* (Davidson 1975) was proposed to carry out large CI matrix calculations. Many approaches for reducing basis functions were also suggested for improving the efficiency of the calculations, e.g., the *pseudopotential* (Heine 1970) approximating inactive orbitals by potentials, the *effective core potential (ECP)* (Kahn and Goddard 1972), replacing core orbitals with potentials, and the *QM/MM method* (Warshell and Levitt 1976) modeling unimportant parts by classical molecular mechanics, are some examples of these approaches. These algorithms and approaches made quantum chemistry calculations practical for investigating the structural and reaction analyses of small molecules.

It is noteworthy that, during this period, DFT was expanded and strengthened in the field of solid state physics. The foundation of DFT was formed by *Janak's theorem* (Janak 1978), the *constrained search formulation* (Levy 1979), the *Runge–Gross theorem* (Runge and Gross 1984) and the requirements for potential functionals were prepared by the *self-interaction correction* (Perdew and Zunger 1981) and so forth. This subsequently led to the explosive growth of potential functionals. Actually, the most frequently used *LDA correlation functional* (Vosko et al. 1980) was developed during this period.

Fifth Stage: Potential Functionals and Excited State Theories (1985–1995)

Science during this period was led by nanomaterials design, inspired by the discovery of C_{60} fullerene in 1985 and by photochemistry inspired by the development of femtosecond time-resolved spectroscopy in 1987. This scientific tide may have had an important impact on quantum chemistry during the latter half of this period: DFT, enabling fast calculations, rapidly grew in use, and many excited state theories were developed for photochemical reaction calculations.

The increasing use of DFT was triggered directly by the development of the *Car–Parrinello molecular dynamic method* (Car and Parrinello 1985), which became the major first-principles molecular dynamics theories based on DFT, and the B88 exchange functional (Becke 1988) and the LYP correlation functional (Lee et al. 1988), which are *GGA functionals* giving accurate results, even for chemical applications. Since quantum chemistry calculations of nanoscale systems were made realistic by these factors, a great number of potential functionals were consequently suggested to obtain higher accuracies in quantum chemistry calculations. The highly popular B3LYP *hybrid functional* (Becke 1993) was also proposed in this period. Moreover, the QM/MM method was applied to DFT for large-scale molecule calculations and resulted in the developments of linear-scaling methods such as the divide-and-conquer method (Yang 1991).

Another trend of quantum chemistry during this period was that various types of *multireference theories* were developed as excited state theories. Single-reference methods such as the SAC-CI method (Nakatsuji and Hirao 1978) had thus far been the mainstream, even in excited state theories. Actually, quantitative discussions of the excited states of small molecules covering femtosecond photochemistry need accurate calculations using multireference theories, which explicitly contain both dynamical and nondynamical electron correlations. Increases in computer performance made it possible to carry out such costly multireference calculations. For instance, major multireference theories, including the CASPT2 (Andersson et al. 1990), MRMP (Hirao 1992), and MCQDPT (Nakano 1993) methods were developed during this period.

Sixth Stage: Easy-to-Use Theories Focusing on Utility (1996–?)

During the last half of the 1990s, theories and technologies in different fields have been integrated, and experimental devices have become highly sophisticated. As a result, various utility theories and technologies have been produced during this period; e.g., human embryonic stem (ES) cells, able to create various human organs (1998), and the optical frequency comb, making it possible to control both the phase and frequency of light (1999). In quantum chemistry, although theories had thus far been specialized for either high speed or high accuracy, utility theories containing both aspects have been required and developed in this period.

In the field of DFT, the *time-dependent response Kohn–Sham method* (Casida 1996), enabling high speed, highly accurate excited state calculations, and *time-dependent current DFT*, for extending the availability of DFT by introducing the vector potential (Vignale and Kohn 1996), were proposed. Then, functionals based on the *long-range correction (LC)* (Iikura et al. 2001), for recovering long-range exchange effects in exchange functionals, and various *semiempirical functionals* (Becke 1997), fitting a huge number of semiempirical parameters to reproduce highly accurate properties, have been produced. The only common characteristic of these functionals is high utility, so as to reproduce accurate results equivalently for a wide variety of property and reaction calculations.

1.2 History of Theoretical Chemistry Prior to the Advent of Quantum Chemistry

Next, let us review the history of theoretical chemistry before the development of the Schrödinger equation (Asimov 1979), because it is also significant to consider the historical orientation of quantum chemistry. This history is basically divided into three stages: genesis, thermal physics–statistical-mechanics stage, and early quantum mechanics stage. Below are brief reviews of each stage.

First Stage: Genesis of Chemistry (–1850s)

Up to the sixteenth century, chemistry was alchemy. Alchemy was based on mysticism originating from Grecian philosophy and expanded in Arabia after ancient Christianity demonized science around 650. At the end of the sixteenth century, A. Libavius changed this circumstance by publishing a chemistry textbook "Alchemy" that avoided mysticism in 1597. R. Boyle then academized chemistry by renaming alchemy as "chemistry" in 1661 and by suggesting the law known as *Boyle's Law*, "pressure times volume is a constant," in 1662.

The foundation of chemistry was constructed by A. de Lavoisier, the "father of modern chemistry." Lavoisier proposed the *law of the conservation of mass* stating "the mass of an isolated system is maintained as a result of processes acting inside the system," and organized the whole knowledge of earlier chemistry in his book, "Traite elementaire de chimie (Elementary Treatise on Chemistry)" (1789). Following the *law of definite composition* (1799) stating "a chemical compound always contains exactly the same proportion of elements by mass," suggested by J.L. Proust, J. Dalton proposed the *law of multiple proportion*, stating "if two elements form more than one compound between them, the ratios of the masses of the second element which combine with a fixed mass of the first element will be ratios of small whole numbers," and called the elements "atoms" for the first time in his book, "A New System of Chemical Philosophy" (1808). He also first suggested the *atomic weight table* in this book. Based on the result of *electrolysis* (1800) by W. Nicholson and A. Carlisle, J.-L. Gay-Lussac proposed the *gas law of combining volumes* (1808), stating "if the mass and pressure of a gas are held constant, then the gas volume increases linearly as the temperature rises," which generalizes the law of multiple proportion. A. Avogadro then suggested the law known as *Avogadro's Law* (1811), stating "under the same conditions of temperature and pressure, equal volumes of all gases contain the same number of molecules." This law made it possible to distinguish *molecules* from atoms. Around the same time, J.J. Berzelius expanded the law of combining volumes to non-integer ratios and suggested *element symbols* and *chemical reaction formulae* (1807–1823), which are still in use today, in his case, to find *isomers* (1830). M. Faraday also elaborated the mechanism of electrolysis and proposed the *law of electrolysis* (1832), which subsequently led to the detection of electrons. Based on the above concepts, E. Frankland set up the ansatz of *valence electrons* (1852) and F.A. Kekule von Stradonitz and A.S. Couper consequently proposed *molecular structural formulae* using *interatomic bonds* (1861), including the *benzene ring* (1865). Thus, the foundations of chemistry were laid.

Second Stage: Thermal and Statistical Mechanics (1840s–1880s)

Around the 1840s, the interest of chemists moved to thermal mechanics. G.H. Hess, the father of thermal mechanics, proved the law known as *Hess's Law* (1840), stating "if a reaction takes place in several steps, then its reaction energy is the sum of the energies of the intermediate reactions into which the overall reaction may be divided at the same temperature." This led to the development of *the first law of thermodynamics* (1842) by J.R. von Mayer, stating "the energy of an isolated system is constant in a thermodynamic process," and *the second law of thermodynamics* (1850) by R. Clausius, stating "the entropy of an isolated system increases in a spontaneous process of energy change." In a study of ether synthesis, A.W. Williamson confirmed the presence of *reversible reactions* and *chemical equilibration*, which led to the development of *chemical reaction kinetics*. Following this study, C.M. Guldberg and P. Waage proposed the *law of mass action*, stating "the reaction rate is proportional to the concentration of matter surrounding the reactant molecules," with *equilibrium equations* (1863).

After the results of basic thermodynamics studies had converged, the thermodynamics of gas molecules began to be understood in terms of the kinetics of constituent molecules. The *kinetic theory of molecules* was launched by J.C. Maxwell, who developed the *velocity distribution function of gas molecules* (1860). L. Boltzmann associated this distribution function with the thermodynamics of gases to propose the *relation between entropy and probability* (1877). By applying a series of thermodynamic laws to chemistry, J.W. Gibbs organized chemical thermodynamics theories by introducing the concepts of *free energy, chemical potential*, and the *phase rule*. Based on these theories, F.W. Ostwald introduced the concept of *catalysis* (1887) and J.H. van 't Hoff suggested the *laws of osmotic pressure* for solutions (1886). S.A. Arrhenius also clarified *ionic dissociations* in electrolyte solutions (1884) and suggested the *activation energies* of reactions (1889). Moreover, the first ever international conference on chemistry was held in Karlsruhe (1860), which induced the classification of the elements and led to the idea of the *periodic table* by D.I. Mendelejev (1869). To fill out the periodic table, elements that included the lanthanides and rare gases were then found one after another. The early period of modern chemistry prior to quantum mechanics became nearly complete in this way.

Third Stage: Early Quantum Mechanics (1890s–1920s)

Beginning in the 1890s, it became the highest priority in chemistry to understand atoms. Although the atomic compositions of molecules and the periodicity of the elements had been clarified, the structures of atoms were difficult to clarify just base on the nature of these properties. The discovery of *electrons* (1897) and the suggestion of the *quantum hypothesis* (1900) at the end of the nineteenth century prefaced the clue to the solution.

The quantum hypothesis was derived by M.K.E.L. Planck from a discussion on the study of black-body radiation. A "black body" is a physical body that absorbs all incident electromagnetic radiation and emits the so-called black-body radiation with a spectrum depending only on temperature. For the spectrum of the black-body radiation, Wien (W.C.W.O.F.F. Wien)'s Law (1886) and the Rayleigh-Jeans

(J.W. Strutt, 3rd Baron Rayleigh, J.H. Jeans) Law (1900) were suggested for high and low frequencies, respectively. However, these laws have problems: the former is inconsistent with classical physics, and the latter yields infinity for the total energy density. Planck proposed a formula giving quite accurate total energies by complementing these laws and advocated the quantum hypothesis, i.e., that this formula is explained by assuming that the energy of each radiation mode is an integral multiple of $h\nu$. Later, the proportionality constant h was called the *Planck constant*. The concept of the *quantum* appeared for the first time in this hypothesis.

The discovery of electrons as particles in a vacuum tube with two metal electrodes (cathode and anode) made by J.J. Thomson also had a large impact on chemistry. In response, P.E.A. von Lenard showed in 1902 that the loss of electrons from metals (and other matter) leads to the photoelectric effect, which is the decrease of electric voltage due to ultraviolet irradiation of the cathode, due to the discharge between the electrodes, which his former supervisor H.R. Hertz had first discovered. However, the photoelectric effect also had an aspect that conflicted with classical electromagnetics. In the photoelectric effect, photoirradiation with a frequency higher than a certain threshold induces current (photocurrent) proportional to the light intensity and electrons (photoelectrons) with energies that were independent of light intensity. Classical electromagnetics can explain this phenomenon. What is unexplainable is the fact that the energies of the electrons increase monotonically with incident light frequency ν.

Einstein knew the quantum hypothesis and advocated the following *photon hypothesis* (Einstein 1905):

- Light is the aggregation of photons, which are energy quanta, $h\nu$.
- Photoabsorption increases the energy of each electrons by $h\nu$.
- Electrons need "work" energies to escape from bulk metals.
- The remaining energy is transformed to the kinetic energy of the electron.

This hypothesis was later proven by the demonstration of the *Compton effect* (Compton 1923).

There was also a controversy concerning the ways electrons exist in atoms. Lenard suggested an atomistic model in which electrons are mixed and paired with positive particles (like positrons) in atoms. Thomson disputed this model because it cannot interpret photoelectric effects and advanced an alternative atomistic model in which negatively charged electrons rotate freely in homogeneous positively charged matter. In response, E. Rutherford considered the experimental result that α particles impinging on metallic foils are scattered with large angles, and proposed an atomistic model, later called the "Rutherford model" (1908), in which electrons circulate around positive charges localized in the center of the atom. A similar atomistic model was also suggested by H. Nagaoka in 1904. Rutherford also proposed that the matter in anode rays, having 1837 times the electron mass, can be used as the fundamental unit of this positive charge (1914). However, even this atomistic model has problems: this model cannot explain the *Rydberg* (J. Rydberg) *formula* (1888), which clearly gives the *emission spectrum of the hydrogen atom*. Moreover, since circulating electrons are undergoing acceleration, classical electromagnetics indicates that electrons in this model must radiate light before eventually falling into the positive charges.

Bohr first proposed an electron motion model to solve the problems of the Rutherford model. For the electron in a hydrogen atom, Bohr presented an atomistic model, in which the periodic orbits of electrons are quantized, and proposed the following hypothesis, known as the "Bohr hypothesis" (Bohr 1913):

- The electron moves in an orbit in which each electron is characterized by a natural number multiplied by the angular momentum $h/2\pi$, and remains in a stationary state without radiating light.
- An electron can be transferred discontinuously from one allowed orbit to another with the absorption or emission of the energy difference, $E - E'$.

This hypothesis leads to the electronic energy of the hydrogen atom and gives the Rydberg formula for the emission spectrum. Moreover, this explains the reason why no electron falls into the positive charge, by assuming that the electrons can exist only in orbits. This atomistic model interpreted the electronic state of the hydrogen atom for the first time. However, various problems remained. This model does not make clear when the electron jumps from one orbit to another and is applicable only to systems having cyclic orbits, like the hydrogen atom. More importantly, *this model cannot describe any electronic state for other atoms containing multiple electrons*. Heisenberg explained that this failure comes from the introduction of classical concepts and symbols and the use of intuitive models and abstractions (Heisenberg 1926). This impasse of early quantum mechanics triggered the paradigm conversion to modern quantum mechanics containing only experimentally verifiable relations.

The key for solving this problem was proposed by L.-V.P.R. de Broglie in his doctoral thesis (1924) 10 years later. De Broglie considered that particles can be regarded as waves, and all matter exists as *matter waves* having a wave-particle duality. That is, all matter has wavelength

$$\lambda = \frac{h}{p}, \tag{1.1}$$

where p is the momentum of the matter. This λ is called the *de Broglie wavelength*. The existence of matter waves explains why the angular momentum of the electron in the hydrogen atom is quantized. Later, the concept of matter waves was confirmed by the Davisson–Germer experiment on electron beam scattering (Davisson and Germer 1927) and the double slit experiment using electron beams (Jönsson 1961). In 1925, Einstein introduced this matter wave study to physicists in Germany, and it consequently led to the development of the Schrödinger equation .

1.3 Analytical Mechanics Underlying the Schrödinger Equation

Before moving on to the Schrödinger equation, let us briefly review the relevant analytical mechanics. The most significant aspects of analytical mechanics are the *least-action principle* and the conservation laws based on it. In 1753, L. Euler arranged P.-L.M. de Maupertuis's thoughts in his paper entitled "On the least-action principle" and proved that the kinetics of mechanical systems obey the least-action principle, to apply this principle to general problems (Ekeland 2009). J.-L. Lagrange proposed his original solution for general problems and named it the *variational method* (1754). Euler introduced this variation method in his paper entitled "Principle of the variation method" (1766) (Ekeland 2009).

The least-action principle leads to the Euler–Lagrange equation determining motion paths. Suppose that a mechanical system is located at two different points with coordinates, q_1 and q_2, at different times, $t = t_1$ and $t = t_2$, respectively. Then, the system transfers between these points under the condition that the action

$$S = \int_{t_1}^{t_2} L(q, \dot{q}, t) dt, \tag{1.2}$$

where q is a general coordinate and $\dot{q}$ is a general velocity, has a stationary value. That is, the variation of action satisfies

$$\delta S = \delta \int_{t_1}^{t_2} L(q, \dot{q}, t) dt \tag{1.3}$$

$$= \frac{\partial L}{\partial \dot{q}} \delta q \bigg|_{t_1}^{t_2} + \int_{t_1}^{t_2} \left(\frac{\partial L}{\partial q} - \frac{d}{dt} \frac{\partial L}{\partial \dot{q}} \right) \delta q dt = 0. \tag{1.4}$$

Application of $\delta q(t_1) = \delta q(t_2) = 0$ yields the *Euler–Lagrange equation* (1766)

$$\frac{\partial L}{\partial q} - \frac{d}{dt} \frac{\partial L}{\partial \dot{q}} = 0, \tag{1.5}$$

where the *Lagrangian* L is defined for independent particles having no interactions between their masses as

$$L = \sum_i \frac{m_i v_i^2}{2} - V(\mathbf{r}_1, \mathbf{r}_2, \ldots). \tag{1.6}$$

For the i-th mass point, m_i is the mass, v_i is the velocity, and $\mathbf{r}_i$ is the position vector. In Eq. (1.6), the first term on the right-hand side is the *kinetic energy*, and V is the *potential energy* which is a function depending only on the positions of the individual masses.

Noether proved a theorem subsequently known as *Noether's theorem*, which assures the conservation laws of energy and momentum from the Euler–Lagrange equation with the uniformities of time and space, respectively (Noether 1918). First, from the uniformity of time, the Lagrangian of independent particle systems does not explicitly depend on time. The total differentiation of the Lagrangian is therefore

$$\frac{dL}{dt} = \sum_i \frac{\partial L}{\partial q_i}\dot{q}_i + \sum_i \frac{\partial L}{\partial \dot{q}_i}\ddot{q}_i. \tag{1.7}$$

Applying the Euler–Lagrange equation in Eq. (1.5) to this equation yields

$$\begin{aligned}\frac{dL}{dt} &= \sum_i \frac{d}{dt}\left(\frac{\partial L}{\partial \dot{q}_i}\right)\dot{q}_i + \sum_i \frac{\partial L}{\partial \dot{q}_i}\ddot{q}_i \\ &= \sum_i \frac{d}{dt}\left(\frac{\partial L}{\partial \dot{q}_i}\dot{q}_i\right).\end{aligned} \tag{1.8}$$

Since this leads to

$$\frac{d}{dt}\left(\sum_i \frac{\partial L}{\partial \dot{q}_i}\dot{q}_i - L\right) = 0, \tag{1.9}$$

it is proven that

$$E = \sum_i \frac{\partial L}{\partial \dot{q}_i}\dot{q}_i - L \tag{1.10}$$

is conserved for the motions of independent particle systems. This E is called the *energy* of independent particle systems. From the uniformity of space, the Lagrangian of independent particle systems remains unchanged for translation in space. For the translation $\mathbf{r}_i \rightarrow \mathbf{r}_i + \Delta\mathbf{r}$ in Cartesian coordinates, the microdisplacement of the Lagrangian is derived as

$$\delta L = \sum_i \frac{\partial L}{\partial \mathbf{r}_i}\Delta\mathbf{r} = \Delta\mathbf{r}\sum_i \frac{\partial L}{\partial \mathbf{r}_i} = 0. \tag{1.11}$$

Since the movement distance $\Delta\mathbf{r}$ is arbitrary, we obtain

$$\sum_i \frac{\partial L}{\partial \mathbf{r}_i} = 0. \tag{1.12}$$

According to the Euler–Lagrange equation in Eq. (1.5), this equation yields

$$\sum_i \frac{d}{dt}\frac{\partial L}{\partial \dot{\mathbf{r}}_i} = \frac{d}{dt}\sum_i \frac{\partial L}{\partial \dot{\mathbf{r}}_i} = 0. \tag{1.13}$$

Therefore, the *momentum* of independent particle systems,

$$\mathbf{p} = \sum_i \frac{\partial L}{\partial \dot{\mathbf{r}}_i}, \tag{1.14}$$

is also conserved. This momentum is described in terms of the general coordinate vector $\mathbf{q}_i$ as

$$\mathbf{p}_i = \frac{\partial L}{\partial \dot{\mathbf{q}}_i}, \tag{1.15}$$

which is the general momentum vector of the i-th particle.

For considering mechanics problems, the formula based on the energy, which is a conserved quantity, is usually superior to that based on the Lagrangian mentioned above. W.R. Hamilton formulated an EOM based on the energy, which excels in its applicability to mechanics problems. With the general momentum p_i, the total differential of the time-independent Lagrangian is given as

$$\begin{aligned} dL &= \sum_i \frac{\partial L}{\partial q_i} dq_i + \sum_i \frac{\partial L}{\partial \dot{q}_i} d\dot{q}_i \\ &= \sum_i \dot{p}_i dq_i + \sum_i p_i d\dot{q}_i . \end{aligned} \tag{1.16}$$

Therefore, applying

$$\sum_i p_i d\dot{q}_i = d\left(\sum_i p_i \dot{q}_i\right) - \sum_i \dot{q}_i dp_i \tag{1.17}$$

to Eq. (1.16) yields

$$d\left(\sum_i p_i \dot{q}_i - L\right) = -\sum_i \dot{p}_i dq_i + \sum_i \dot{q}_i dp_i . \tag{1.18}$$

The representation in parentheses on the left-hand side is a formula for the energy, which is called the *Hamiltonian* of the system. Equation (1.18) leads to the *canonical equation*,

$$\dot{q}_i = \frac{\partial H}{\partial p_i}, \quad \dot{p}_i = -\frac{\partial H}{\partial q_i}. \tag{1.19}$$

Thus far, the action in Eq. (1.4) has been used as an auxiliary quantity for representing the EOM. Let us redefine the action as a physical quantity characterizing the least-action motion. Given $\delta q(t_1) = 0$ and $\delta q(t_2) = \delta q$, the Euler–Lagrange equation in Eq. (1.5) and $p_i = \partial L / \partial \dot{q}_i$ leads to the variation of the action in Eq. (1.4),

$$\delta S = \sum_i p_i \delta q_i. \tag{1.20}$$

It follows that

$$\frac{\partial S}{\partial q_i} = p_i. \tag{1.21}$$

From the definition of the action, the total differential of the action in terms of time can be written as

$$\frac{dS}{dt} = L. \tag{1.22}$$

Assuming that the action is a function of coordinates and time, the Lagrangian is given as

$$L = \frac{dS}{dt} = \frac{\partial S}{\partial t} + \sum_i \frac{\partial S}{\partial q_i}\dot{q}_i = \frac{\partial S}{\partial t} + \sum_i p_i \dot{q}_i. \tag{1.23}$$

From the definition of the Hamiltonian in Eq. (1.18), we therefore find

$$\frac{\partial S}{\partial t} = L - \sum_i p_i \dot{q}_i = -H. \tag{1.24}$$

Since the Hamiltonian is a function of the coordinates, momenta, and time, and the momentum is given in Eq. (1.14), we obtain

$$\frac{\partial S}{\partial t} = -H(q, p, t) = -H\left(q, \frac{\partial S}{\partial q}, t\right), \tag{1.25}$$

which is called the *(time-dependent) Hamilton–Jacobi equation.* Provided that the Hamiltonian does not explicitly depend on time, Eq. (1.10) leads to

$$L = \sum_i p_i \dot{q}_i - E. \tag{1.26}$$

Since the action can be separated into a factor depending only on the coordinates and a factor explicitly depending on time, it follows that

$$S = S_0(q) - Et. \tag{1.27}$$

Substituting this into the Hamilton–Jacobi equation gives

$$H\left(q, \frac{\partial S}{\partial q}\right) = E, \tag{1.28}$$

which is called the *time-independent Hamilton–Jacobi equation* or the *energy conservation equation*. These Hamilton–Jacobi equations were used to develop the Schrödinger equation.

The above equations of motion determining the motions of mechanical systems collectively make up what is called *analytical mechanics*. The phrase "analytical" was added to distinguish it from the unique mechanics using geometrical configurations seen in the "Philosophia Naturalis Principia Mathematica" (1687) by I. Newton. Note that it is usually difficult to solve the EOM for specific mechanical systems. Solving the EOM implies that one can predict the state at a given time in the future from the initial state. For most mechanical model systems, we cannot analytically solve the EOM. J.-H. Poincaré proved that *there are very few classical mechanics problems, for which the EOM can be solved analytically* (Ekeland 2009). Even in the present day, the EOM is generally solved by approximate methods. Since no computers were available in the age of Poincaré, he confined these systems to those with periodic solutions. The least-action principle underlying the EOM should be renamed the "stationary"-action principle, because the action is not exclusively the least action (Ekeland 2009). For the "stationary"-action principle, Feynman made this clear later with the idea of the *path integral* based on quantum mechanics (Feynman 1948). That is, the *classical paths of the stationary action are only the paths of overwhelmingly higher probability compared to others*. Theoretical calculation results based on the path integral agree with experimental values with quite high accuracy. Furthermore, the question still remains as to how to determine whether the EOM has a solution or not. This problem is one of "Hilbert's 23 problems." In the present day, it can be determined whether a solution exists for a given system or not (Ekeland 2009).

1.4 Schrödinger Equation

In 1926, Schrödinger published the first paper of "Quantisierung als Eigenwert problem (Quantization as an Eigenvalue Problem)" (Schrödinger 1926). In this paper, Schrödinger proposed a new equation combining de Broglie's concept of matter waves with the Hamilton–Jacobi equation, the *Schrödinger equation*. Later, Dirac (Ph.D. thesis, 1926) and Jordan (Born et al. 1926) independently proved that this equation is identical to the matrix equation suggested by Heisenberg (1925).

Schrödinger proved that a normal but somewhat mysterious quantization rule is naturally provided by assuming the finiteness and definiteness of a spatial function Ψ (Schrödinger 1926). Assuming that Ψ is a logarithmic function representing action S, a sum of functions, as a product, he defined

$$S = i\hbar \ln \Psi, \tag{1.29}$$

where Ψ is given as

$$\Psi = \exp\left(-\frac{iS}{\hbar}\right). \tag{1.30}$$

Since this Ψ is regarded as the amplitude of matter waves, *the finiteness and definiteness of Ψ can be presumed by a normalization condition*,

$$\int |\Psi|^2 \, d\tau = 1. \tag{1.31}$$

This Ψ is called a *wavefunction*. For an energy conservative system having an action in Eq. (1.27), the variables of this Ψ can be separated into

$$\Psi = \Psi_0 \exp\left(-\frac{iEt}{\hbar}\right), \quad \Psi_0 = \exp\left(-\frac{iS_0}{\hbar}\right). \tag{1.32}$$

Substituting the action S in Eq. (1.29), using this wavefunction Ψ, into the time-dependent and time-independent Hamilton–Jacobi equations in Eqs. (1.25) and (1.28) gives

$$\hat{H}\Psi = i\hbar\frac{\partial \Psi}{\partial t} \tag{1.33}$$

and

$$\hat{H}\Psi = E\Psi, \tag{1.34}$$

respectively. Equations (1.33) and (1.34) are called the *time-dependent and time-independent Schrödinger equations*, respectively. Note that the Hamiltonian is replaced with the Hamiltonian operator $\hat{H}$ acting on spatial functions.

As an example, let us consider a time-independent, independent-particle system. Since Eq. (1.29) leads to

$$\begin{aligned} \hat{H} &= \sum_i \frac{p_i^2}{2m_i} + V = \sum_i \frac{1}{2m_i}\left(\frac{\partial S}{\partial q_i}\right)^2 + V \\ &= -\sum_i \frac{\hbar^2}{2m_i}\frac{1}{|\Psi|^2}\left(\frac{\partial \Psi}{\partial q_i}\right)^2 + V = E, \end{aligned} \tag{1.35}$$

we obtain an energy conservation equation,

$$-\sum_i \frac{\hbar^2}{2m_i}\left(\frac{\partial \Psi}{\partial q_i}\right)^2 + (V - E)|\Psi|^2 = 0. \tag{1.36}$$

The stationary condition for this equation under the normalization condition of Ψ is derived using the variational method as

$$-\sum_i \frac{\hbar^2}{2m_i}\left(\frac{\partial^2 \Psi}{\partial q_i^2}\right) + (V - E)\Psi = 0. \quad (1.37)$$

For a three-dimensional system, this equation is given as the familiar Schrödinger equation,

$$-\sum_i \frac{\hbar^2}{2m_i}\nabla_i^2 \Psi + (V - E)\Psi = 0. \quad (1.38)$$

What should be noted is that the momentum $\mathbf{p}_i$ is transformed to a quantum operator,

$$\mathbf{p}_i = -i\hbar\nabla_i. \quad (1.39)$$

This momentum operator produces discrete energy levels under the boundary condition of potential V to make the wavefunctions Ψ finite. The discrete energy values and the corresponding wavefunctions are called *eigenvalues* and *eigenfunctions*, respectively.

1.5 Interpretation of the Wavefunction

Since the Schrödinger equation was developed, the interpretation of the wavefunction has been vigorously discussed. Schrödinger himself suggested a *wave interpretation* for the wavefunction (Jammer 1974). From this interpretation, the physical existence of a matter consists only of waves, and the discrete eigenvalues are not the energies but the eigenfrequencies of the waves. He also insisted that it is meaningless to assume discrete energy levels and quantum transitions independently as seen in matrix mechanics. However, this wave interpretation has many serious problems. Although a particle is taken to be a *wave packet* in this interpretation, this indicates that the wave packet has to be a series expansion of the integral multiples of normal vibrations. This wave packet is applicable only to the wavefunctions for harmonic oscillators (see Sect. 1.7). This interpretation is also available only for three-dimensional cases, even though the motions of n particle systems require $3n$ dimensions. Moreover, this interpretation assumes that the wavefunction is real, but this conflicts with the fact that the wavefunction is complex. Note, however, that there is no practical problem if the Hamiltonian contains no vector potential (see Sect. 6.5). There are various other problems, e.g., this interpretation cannot explain why the wavefunction changes discontinuously due to a measurement.

As an alternative for the wave interpretation, Born proposed a *probabilistic interpretation* (Jammer 1974). According to this interpretation, which originates from the discussion of the quantum treatment of collision processes, wavefunction-based mechanics targets only the existence probability P of particles in a differential volume element $d\tau$,

$$P(\tau)d\tau = |\Psi|^2 d\tau, \tag{1.40}$$

and this probability is supposed to act in a classical way. Based on this interpretation, the expectation value of a function f is calculated as

$$\langle f \rangle = \int d\tau f(\tau) P(\tau) = \int d\tau \Psi^*(\tau) f(\tau) \Psi(\tau). \tag{1.41}$$

This interpretation proposes that the *wavefunction indicates neither a physical system nor its physical attributes but information on the latter.* A common misunderstanding is that this interpretation suggests that the wavefunction is an existence probability. In this case, it is unaccountable why double slit experiments produce spots of particles. Actually, the probabilistic interpretation denies this explanation and states that the wavefunction is something other than the existence probability. This interpretation only focuses attention on the existence probability in discussions on the wave-particle relation. Eventually, *the Copenhagen school, led by Bohr, Heisenberg and Pauli, advocated Born's probabilistic interpretation, and the latter has become the mainstream interpretation of the wavefunction at present.* However, this probability interpretation has not solved the interpretation problem of the wavefunction. The cause is an interpretation added to the probabilistic interpretation by the Copenhagen school: *phenomena are accountable only by probabilities, and wave packets are reduced by observations.* To counter this additional interpretation, Einstein advocated the *hidden-variable interpretation*, i.e., that a hidden variable makes quantum mechanics accountable only by probabilities. The famous phrase "God doesn't play dice" emerged out of this context. Consequently, the so-called Bohr–Einstein debates were conducted in discussions at Solvay conferences and correspondences, in which Bohr argued against Einstein's counterarguments (Jammer 1974). Einstein's main counterarguments, based on thought experiments, and Bohr's refutations, are summarized as follows:

- Reduced wave packets are statistical distribution functions taken to be real. ⇒ Observations transform systems in the instance that wave packets are reduced. Therefore, this argument is meaningless.
- In a *double-slit experiment*, if the kick of a momentum to the slit is measured when a particle passes through one side of the double slit, the path of the particle can be determined without measuring the particle itself. ⇒ This setup of the experiment shifts the quantum state of the particle.
- The energy of a photon can be determined in a *photon-box experiment* by measuring the energy of a photon box after emission of a photon. ⇒ This conflicts with the uncertainty relation between energy and time.

- In the case that a particle with spin 0 decays into two electrons, observing the spin of one electron determines the spin of the second electron. Since this indicates that information transmits faster than light, it violates the relativistic theory (*Einstein–Podolsky–Rosen paradox*) (Einstein et al. 1935). ⇒ Bohr could not provide a counterargument to this. Later, this paradox was resolved by *Bell's inequality*, which limits the correlation of subsequent measurements of particles that have interacted and then separated on the local hidden-variable theory (Bell 1964), and *Aspect's experiment*, which proves the violation of this inequality (Aspect et al. 1982).

It is widely believed that von Neumann's *no-go theorem* (1932) decided the overall outcome in the above debates. This theorem given in "Die Mathematische Grundlagen der Quantenmechanik (Mathematical Foundations of Quantum mechanics)" (von Neumann 1957) mathematically proves that *the Schrödinger equation contains no hidden variables.* Although this theorem nearly led to the end of the debates, Schrödinger was not satisfied and countered with a thought experiment called *Schrödinger's cat* (Schrödinger 1935). Suppose a box containing a cat and a Geiger counter linked to a cyanide-gas generator when sensing an α particle from a radioactive isotope. Based on the probabilistic interpretation, after a period of time, the cat should be situated at the superposition of life and death until getting the box is opened, despite the fact that it is actually one or the other. Schrödinger declared this to be paradoxical. Although this thought experiment cannot disprove the no-go theorem, it subsequently led to *Everett's many-worlds interpretation* (Everett 1957), in which observations do not reduce wave packets but bifurcate the world. Furthermore, Bohm, who was once an assistant professor under Einstein, indicated that the prerequisite in the proof of the no-go theorem is too strict to be general, and derived the classical mechanics from the Schrödinger equation by introducing "quantum-mechanical force" as a hidden variable (Bohm 1952). However, this interpretation was also denied by Aspect's experiment (Aspect et al. 1982) and was disproved by the *Kochen–Specker theorem* (Kochen and Specker 1967). The interpretation of wavefunction is still being debated. Actually, many studies are still carried out to suggest a quantum mechanics based on a delocalized hidden variable, because the Kochen–Specker theorem disproves only the existence of localized hidden variable.

1.6 Molecular Translational Motion

Next, let us consider the eigenstates of molecular motions in the Schrodinger equation. *The molecular motions are classified into four types: the translational, rotational, and vibrational motions of atomic nuclei* (Fig. 1.1) *and the motions of electrons.* The translational motions are the uniform motions of all nuclei with three degrees of freedom (DOFs), the rotational motions are those with respect to the centroids of molecules, with three DOFs (two DOFs for linear molecules), and

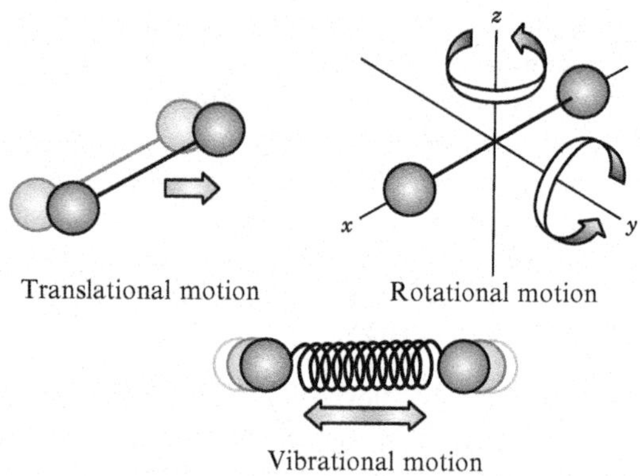

Fig. 1.1 Three basic motions of atomic nuclei in molecules

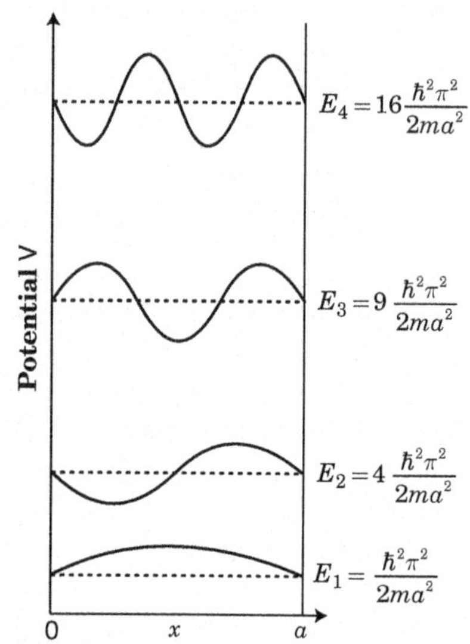

Fig. 1.2 One-dimensional box potential with the width of a and the corresponding energy eigenvalues, E_i ($i = 1, 2, \ldots$), and the images of translational eigenfunctions, Ψ_i. In the eigenvalues, m is the mass of the particle in the potential and $\hbar$ is the reduced Planck constant

the vibrational motions are periodic motions centering on the equilibrium structures of molecules with $3N-6$ DOFs ($3N-5$ DOFs for linear molecules). In this chapter, the quantum eigenstates of these motions are specifically viewed henceforth.

First, since the translational motion of molecules can be assumed to be the particle motion of the centroid, it is usually taken as the simplest elementary problem: *the box potential problem*. Concerning the method of solving this problem, readers may easily learn about this in an elementary book on quantum mechanics. Now, let us consider the eigenstates of the translational motions of molecules in the box potential. For the simplest one-dimensional box potential with width a (Fig. 1.2), the energy eigenvalues are determined as

$$E_n = \frac{\hbar^2 k^2}{2m} = \frac{\hbar^2 \pi^2 n^2}{2ma^2} \quad (n = 1, 2, 3, \ldots), \tag{1.42}$$

where m is the mass of the molecule. The corresponding normalized eigenfunctions are obtained as

$$\Psi_n(x) = \left(\frac{2}{a}\right)^{1/2} \sin(kx). \tag{1.43}$$

These eigenvalues and eigenfunctions have the following characteristics (Gasiorowicz 1996).

- The energy of the lowest eigenstate, called the *ground state*, is not zero but has the value

$$E_1 = \frac{\hbar^2\pi^2}{2ma^2}. \tag{1.44}$$

  This is in contrast to the classical state of a particle at rest in a hole, for which the sum of the kinetic energy and potential energy is zero. This nonzero energy is called the *zero-point energy*.
- Given the momentum of a molecule, the expectation value of the momentum $\langle p \rangle$ is always zero for real eigenfunctions. This is because for an arbitrary real function $R(x)$,

$$\int dx R(x)(-i\hbar)\frac{dR}{dx}(x) \tag{1.45}$$

  is always imaginary and therefore $\langle p \rangle$ is not identical to the complex conjugate $\langle p \rangle^*$ except for $\langle p \rangle = 0$.
- The increase in the number of *nodes* of an eigenfunction causes the energy eigenvalue to increase, because the kinetic energy,

$$\langle T \rangle = -\frac{\hbar^2}{2m}\int dx \Psi^*(x)\frac{d^2\Psi}{dx^2}(x), \tag{1.46}$$

  increases as the curvature of the eigenfunction grows.
- Since the eigenfunctions are *unit vectors*, an arbitrary function is representable as an eigenfunction expansion. This is because, based on Fourier's theorem that an arbitrary function can be expanded by the series expansion of trigonometrical functions, a function is always expanded by the eigenfunctions of the translational motions, which are trigonometrical functions.

Although the box potential has so far been considered, the potential-free ($V = 0$) case may be close to the readers' image of translational motion. Such potential-free particles are called *free particles*. In the one-dimensional case, the eigenfunctions of free particles are given by

$$\Psi(x) = \exp(\pm ikx) \text{ or } \Psi(x) = \{\sin(kx), \cos(kx)\} \tag{1.47}$$

and its linear combinations. These eigenfunctions are characterized as follows (Gasiorowicz 1996):

- For the momentum operator $\hat{p} = -i\hbar(d/dx)$, the eigenfunctions in Eq. (1.47) give the momentum eigenvalue k and the corresponding energy eigenvalue $k^2/2m$. Since the momentum k is a real number, the energy eigenvalues are continuous. The eigenstates giving continuous energy eigenvalues are called *continuum states*. For the continuum states, the eigenfunctions, which are always finite, give a quite large overlap integral with an eigenfunction with a different momentum k' specifically at $k = k'$,

$$\int_{-\infty}^{\infty} dx\,(\exp(\pm ikx))^* \exp(\pm ik'x) = 2\pi\delta(k - k'). \tag{1.48}$$

  The function $\delta(k - k')$, called the *delta function*, gives a quite large value at $k = k'$.
- The eigenfunctions in Eq. (1.47) have two forms giving the same energy eigenvalue. This situation is termed *degeneracy*, i.e., when there are two or more independent eigenfunctions corresponding to the same eigenvalue. Since degenerate eigenfunctions are orthogonal to each other, the linear combination of these eigenfunctions is also an eigenfunction. The degeneracy results from the existence of another operator which is commutable (commutative) with the operator giving the eigensolution (the Hamiltonian in the present case). In this case, the commutative parameter is the momentum operator $\hat{p}$ and it gives a different eigenvalue,

$$\hat{p}\exp(\pm ikx) = -i\hbar\frac{d}{dx}\exp(\pm ikx) = \pm k\hbar\exp(\pm ikx), \tag{1.49}$$

  for each eigenfunction.
- In general, *an operator depending explicitly on time is not commutative with the Hamiltonian, and therefore it does not conserve energy*, i.e., energy is not conserved for a potential $V(x, t)$ that is an explicit function of time. This is because the energy conservation law is based on the uniformity of time, as shown in Sect. 1.3. In this case, no degenerate energy eigenvalue is therefore provided. One of the operators that does not explicitly depend on time and is commutative with the Hamiltonian is the *parity operator*, which reverses the coordinate axis. In the eigenfunctions of Eq. (1.47), $\sin(kx)$ and $\cos(kx)$ are degenerate for the parity operator and give different eigenvalues, -1 and $+1$, respectively.

1.7 Molecular Vibrational Motion

Next, let us consider the vibrational motions of molecules. The simplest model of the vibrational motion is the *harmonic oscillator*. The harmonic oscillator, an ideal spring motion, is represented as a potential $V = k\mathbf{q}^2/2$, in which k is the spring constant, and $\mathbf{q}$ is a coordinate vector centered on the equilibrium structure. For the one-dimensional spring motion on the x axis, the energy eigenvalues of the Schrödinger equation for the harmonic oscillator potential are given as

$$E_v = \left(v + \frac{1}{2}\right)\hbar\left(\frac{k}{\mu}\right)^{1/2} = \left(v + \frac{1}{2}\right)\hbar\omega \ (v = 0, 1, 2, \ldots). \quad (1.50)$$

In Eq. (1.50), the reduced mass μ is calculated as $\mu = m_1 m_2/(m_1 + m_2)$ for the spring motion of the two masses m_1 and m_2. Despite vibrational motions being one-dimensional, ω is called the "angular" frequency, because vibrational motions periodically depend on time, as in $\exp(i\omega t)$. It is interesting to note that the differences in the energy eigenvalues are equal to integral multiples of $\hbar\omega = h\nu$ and therefore support Planck's hypothesis concerning black-body radiation. The eigenfunctions corresponding to these eigenvalues are given by

$$\Psi^{\mathrm{vib}}(y) = h(y)\exp(-y^2/2) \quad (1.51)$$

$$y = \left(\frac{\mu\omega}{\hbar}\right)^{1/2} x, \quad (1.52)$$

where h is a function satisfying

$$\frac{d^2h}{dy^2}(y) - 2y\frac{dh}{dy}(y) + (\epsilon - 1)h(y) = 0, \quad (1.53)$$

which is identical to the Hermite polynomial,

$$H_n(y) = (-1)^n \exp(y^2)\frac{d^n}{dy^n}\exp(-y^2), \quad (1.54)$$

without the normalization factor (see Fig. 1.3). For the vibrational motions, the energy eigenvalues and eigenfunctions have characteristics similar to those of the translational motions. Let us consider the simplest harmonic oscillator model. The lowest energy ($n = 0$) is not zero, as shown in Eq. (1.50) and is called the *zero-point vibrational energy*. With increasing number of nodes of the eigenfunctions, the energy eigenvalues increase. Moreover, the expectation value of momentum is confirmed to be zero by Eq. (1.45), and the eigenfunctions can be taken as unit vectors.

Although the harmonic oscillator model is an appropriate approximation for the vibrational motions of molecules, despite its simplicity, we should bear in mind that the actual potentials of molecular vibrations are dissimilar to the harmonic oscillator for high-level eigenstates. Even for the bonds of diatomic molecules, the potentials of the vibrational motions are in the form of the *Morse potential* (Fig. 1.4), which significantly differs from the harmonic oscillator potential due to the anharmonicity associated with large internuclear distances. For this potential, the energy eigenvalues are given as

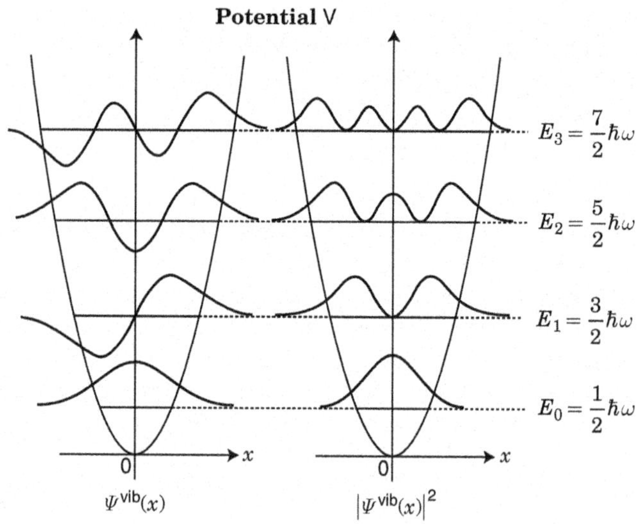

Fig. 1.3 Harmonic oscillator potentials and corresponding energy eigenvalues, E_i ($i = 0, 1, 2, \ldots$), vibrational eigenfunctions, Ψ^{vib}, (*left*) and existence probability, $|\Psi^{\mathrm{vib}}|^2$, (*right*). In the eigenvalues, ω is the angular frequency

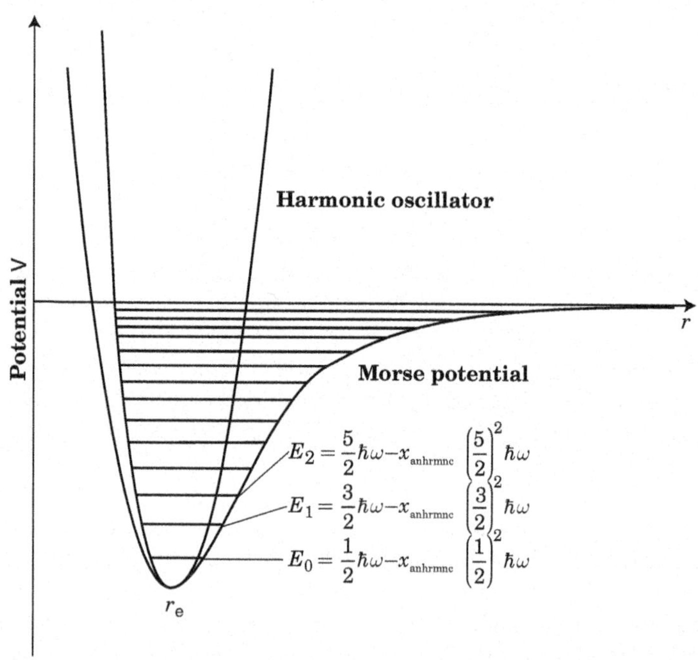

Fig. 1.4 Morse potential, compared to harmonic oscillator potential, and its energy eigenvalues, E_i ($i = 0, 1, 2, \ldots$). In the figure, x_{anhrmnc} is the anharmonic constant and r_e is the equilibrium distance

$$E_v = \left(v + \frac{1}{2}\right)\hbar\omega - x_{\mathrm{anhrmnc}}\left(v + \frac{1}{2}\right)^2 \hbar\omega \quad (v = 0, 1, 2, \ldots), \tag{1.55}$$

where x_{anhrmnc} is termed the anharmonic constant.

Finally, let us consider the relationship between the vibrational motions and the infrared (IR) absorption spectra. The IR spectra show the frequencies corresponding to the energy gaps in the transitions between vibrational eigenstates, with the peak intensities proportional to the transition moments. The transitions between vibrational states have rules called selection principles: for the harmonic oscillator, transitions take place for the eigenstate pairs with $\Delta n = \pm 1$. This selection principle comes from the fact that the transition moment, which is proportional to the transition dipole moment,

$$\mu_{ij} = \int d^3\mathbf{r}\Psi_i^{\mathrm{vib}}\mathbf{p}_{\mathrm{ed}}\Psi_j^{\mathrm{vib}}, \tag{1.56}$$

is zero for even values of $|i - j|$ and nearly zero for odd values of $|i - j| \geq 3$, except for $j = i \pm 1$. In Eq. (1.56), $\mathbf{p}_{\mathrm{ed}}$ is the electric dipole moment, which is proportional to the expectation value of the position vectors of the electrons from the centroid of the atomic nuclei. *At room temperature, since most molecules are in their vibrational ground states, the spectral peaks are usually assigned to the transition* $v = 0 \to 1$, for which the spectral peaks are termed elementary bands. The spectral peaks corresponding to the transitions $v = 1 \to 2$, which increasing in importance with rising temperature (termed "hot bands") are indistinguishable for the harmonic oscillator. However, the large anharmonicity in vibrational potentials makes the hot bands distinguishable and allows the overtone transitions corresponding to $v = 0 \to 2$, violating the selection principle.

1.8 Molecular Rotational Motion

Regarding the motions of atomic nuclei, those that remain are the rotational motions. In this section, rotational motion is explained in somewhat more detail, because it is concerned with the nature of chemistry. What is important to consider is that the energies associated with the rotational motions of molecules are part of the kinetic energies, and therefore overlap with the translational motion energies without operation. That is, a variable separation is required for the translational motions, which are the motions of entire systems, and the rotational motions, which are internal motions (Gasiorowicz 1996). In the case of diatomic molecules, the Hamiltonian operator is given as

$$\hat{H} = \frac{\hat{\mathbf{p}}_1^2}{2m_1} + \frac{\hat{\mathbf{p}}_2^2}{2m_2} + V(r), \tag{1.57}$$

where m_1 and m_2 are the masses of two atoms, $\hat{\mathbf{p}}_1$ and $\hat{\mathbf{p}}_2$ are the momentum vectors of these atoms, and r is the interatomic distance. The variable transformation can be done for this Hamiltonian operator as

$$\hat{H} = \frac{\hat{\mathbf{P}}^2}{2M} + \frac{\hat{\mathbf{p}}^2}{2\mu} + V(r), \tag{1.58}$$

where the total momentum $\hat{\mathbf{P}}$ and relative momentum $\hat{\mathbf{p}}$ are given by

$$\hat{\mathbf{P}} = \hat{\mathbf{p}}_1 + \hat{\mathbf{p}}_2, \quad \hat{\mathbf{p}} = \frac{m_2\hat{\mathbf{p}}_1 - m_1\hat{\mathbf{p}}_2}{m_1 + m_2}, \tag{1.59}$$

respectively. In Eq. (1.59), $M = m_1 + m_2$ is the total mass and $\mu = m_1 m_2 / M$ is the reduced mass. By this variable transformation, the rotational motion can be taken as the central force motion of a body with reduced mass μ. Since the total momentum operator $\mathbf{P}$ is commutative with $\hat{H}$, the wavefunction can be separated into translational and rotational motion terms such as

$$\Psi(\mathbf{R}, \mathbf{r}) = \Psi_{\mathbf{P}}(\mathbf{R})\Psi_{\mathbf{p}}(\mathbf{r}), \tag{1.60}$$

where

$$\Psi_{\mathbf{P}}(\mathbf{R}) = (2\pi\hbar)^{-3/2} \exp(i\hat{\mathbf{P}} \cdot \mathbf{R}/\hbar). \tag{1.61}$$

The Schrödinger equation only for the rotational motions is therefore provided as

$$\hat{H}\Psi_{\mathbf{p}}(\mathbf{r}) = \left(\frac{\hat{\mathbf{p}}^2}{2\mu} + V(r)\right)\Psi_{\mathbf{p}}(\mathbf{r}) = E\Psi_{\mathbf{p}}(\mathbf{r}). \tag{1.62}$$

The Hamiltonian operator is invariant for rotations because of the energy conservation of the rotational motion. With an infinitesimal rotation around the z axis,

$$\begin{pmatrix} x' \\ y' \end{pmatrix} = \begin{pmatrix} \cos\theta & -\sin\theta \\ \sin\theta & \cos\theta \end{pmatrix} \begin{pmatrix} x \\ y \end{pmatrix} \simeq \begin{pmatrix} 1 & -\theta \\ \theta & 1 \end{pmatrix} \begin{pmatrix} x \\ y \end{pmatrix}, \tag{1.63}$$

the Schrödinger equation in Eq. (1.62) is transformed to

$$\hat{H}\Psi_{\mathbf{p}}(x - \theta y, y + \theta x, z) = E\Psi_{\mathbf{p}}(x - \theta y, y + \theta x, z). \tag{1.64}$$

By expanding this equation in terms of θ, the first-order term is given as

$$\hat{H}\left(x\frac{\partial}{\partial y}-y\frac{\partial}{\partial x}\right)\Psi_{\mathbf{p}}(x,y,z)=E\left(x\frac{\partial}{\partial y}-y\frac{\partial}{\partial x}\right)\Psi_{\mathbf{p}}(x,y,z)$$
$$=\left(x\frac{\partial}{\partial y}-y\frac{\partial}{\partial x}\right)\hat{H}\Psi_{\mathbf{p}}(x,y,z). \quad (1.65)$$

Defining the rotational operator around the z axis,

$$\hat{L}_z=\frac{\hbar}{i}\left(x\frac{\partial}{\partial y}-y\frac{\partial}{\partial x}\right)=xp_y-yp_x, \quad (1.66)$$

the relation between $\hat{H}$ and $\hat{L}_z$,

$$\left(\hat{H}\hat{L}_z-\hat{L}_z\hat{H}\right)\Psi_{\mathbf{p}}(x,y,z)=0 \quad (1.67)$$

is obtained. This indicates that $\hat{H}$ and $\hat{L}_z$ are commutative. Similarly, $\hat{H}$ and $\hat{L}_x$, $\hat{H}$ and $\hat{L}_y$ are also commutative. Although $\hat{L}_x$, $\hat{L}_y$, and $\hat{L}_z$ are noncommutative with each other, these operators are commutative with $\hat{\mathbf{L}}^2=\hat{L}_x^2+\hat{L}_y^2+\hat{L}_z^2$. Therefore, $\hat{H}$, $\hat{L}_z$, and $\hat{\mathbf{L}}^2$ are commutative and have simultaneous eigenfunctions. This is a characteristic of the rotational motion. Using $\mathbf{r}=(x,y,z)$ and $\mathbf{p}=-i\hbar(\partial/\partial x,\partial/\partial y,\partial/\partial z)$, the angular momentum can be written as

$$\hat{\mathbf{L}}^2+(\mathbf{r}\cdot\hat{\mathbf{p}})^2=r^2\hat{\mathbf{p}}^2+i\hbar\mathbf{r}\cdot\hat{\mathbf{p}}. \quad (1.68)$$

The square of the momentum is therefore

$$\hat{\mathbf{p}}^2=\frac{1}{r^2}\left(\hat{\mathbf{L}}^2+(\mathbf{r}\cdot\hat{\mathbf{p}})^2-i\hbar\mathbf{r}\cdot\hat{\mathbf{p}}\right)=\frac{1}{r^2}\hat{\mathbf{L}}^2-\hbar^2\frac{1}{r^2}\left(r\frac{\partial}{\partial r}\right)^2-\hbar^2\frac{1}{r}\frac{\partial}{\partial r}. \quad (1.69)$$

Since this reduces the angular terms of the Schrödinger equation represented in polar coordinates (r,θ,ϕ) to the terms of $\mathbf{L}^2$, the eigenfunctions can be separated into the angular and radial functions,

$$\Psi_{\mathbf{p}}(\mathbf{r})=Y_{\lambda}(\theta,\phi)R_{\mathbf{p}\lambda}(\mathbf{r}). \quad (1.70)$$

This angular function Y_λ is called the *spherical harmonic function*. Since the spherical harmonic function is also an eigenfunction of the rotational operator around the z axis, L_z, it can be further separated into functions of the angles θ and ϕ. Defining the rotational angle around the z axis as ϕ, the rotation operator around the z axis is represented as $L_z=-i\hbar\partial/\partial\phi$. The eigenequations and eigenfunctions of L_z are given as

$$\frac{\partial\Phi_{m_l}}{\partial\phi}=im_l\Phi_{m_l}(\phi), \quad (1.71)$$

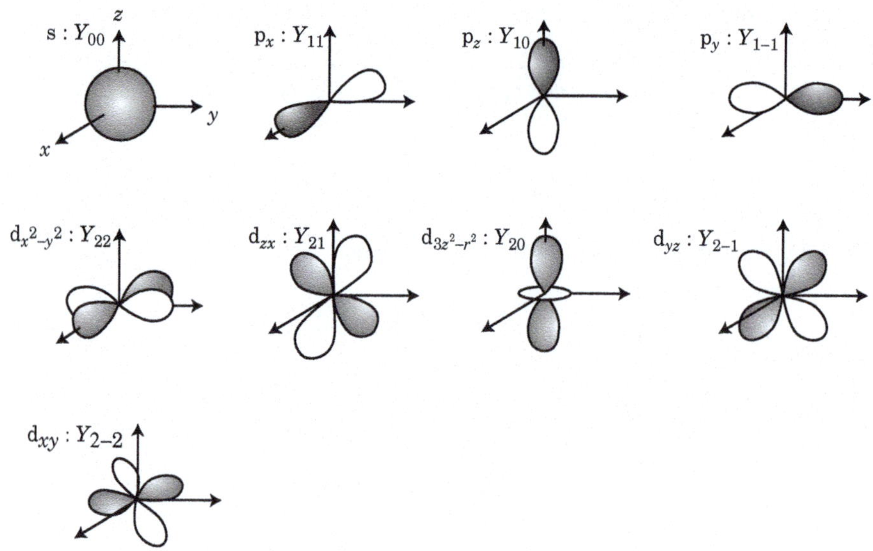

Fig. 1.5 Spherical harmonic functions, Y_{lm_l}, where l and m_l are angular momentum and magnetic quantum numbers

and

$$\Phi_{m_l}(\phi) = \frac{1}{(2\pi)^{1/2}} \exp(i m_l \phi), \tag{1.72}$$

where m_l is called the *magnetic quantum number* and has integral values ($m_l = 0, \pm 1, \pm 2, \ldots$). The remaining eigenfunction for θ is the eigenfunction of $\mathbf{L}^2$, which is derived to be the solution of the Legendre equation,

$$\frac{d}{dz}\left[(1-z^2)\frac{\partial^2 \Theta}{\partial z^2}(z)\right] + \left(\frac{\lambda}{\hbar^2} - \frac{m_l{}^2}{1-z^2}\right)\Theta(z) = 0. \tag{1.73}$$

This eigenfunction Θ has a solution only for $\lambda = l(l+1)\hbar^2$ (l is called the *azimuthal quantum number*, which is a natural number). This λ is also the eigenvalue of $\mathbf{L}^2$ of the spherical harmonic function. Since the magnetic quantum numbers m_l are restricted to $|m_l| \leq l$, i.e. $-l \leq m_l \leq l$, *the eigenstates are* $(2l+1)$-*fold degenerate.* Therefore, the spherical harmonic function is specified as Y_{lm_l} (Fig. 1.5). For the remaining radial function, it is complicated to determine a specific form, because the potential V significantly depends on the types of interatomic bonds in the rotational motions of the molecules. However, the quantum nature of the radial functions is fortunately negligible, because the interatomic bond potentials are extremely deep in general. It is therefore reasonable to consider that the eigenfunctions of the rotational motions are the spherical harmonic functions in most cases. Since the Hamiltonian operator, neglecting the radial part, is

$$\hat{H} = \frac{\mathbf{L}^2}{2\mu r^2}, \tag{1.74}$$

the energy eigenvalues for the rotational motions are given by

$$E = \frac{\mathbf{L}^2}{2\mu r^2} = \frac{l(l+1)\hbar^2}{2I}, \tag{1.75}$$

where $I = \mu r^2$ is the moment of inertia.

1.9 Electronic Motion in the Hydrogen Atom

While the quantizations of nuclear motions in molecules have been thus far reviewed, the eigenstates for the electronic motions in molecules are also determined by solving the Schrödinger equation. However, *the eigenstates of the electronic motions cannot be analytically determined except for extremely simple systems such as the hydrogen atom and hydrogen-like atoms*. The cause and solution for this are described starting in the next chapter. In this section, let us consider the eigenstate of the electronic motion in the hydrogen atom, which is one of the simplest systems.

In the hydrogen atom, the electron is assumed to move in a circle around the nucleus, attracted by the Coulomb electrostatic force. The eigenstate of the electronic motion is therefore determined by the Schrödinger equation for the rotational motion. The Hamiltonian operator is in the same form as that for the rotational motions of molecules mentioned in the last section. Following Eq. (1.57), the Hamiltonian operator is represented as

$$H = \frac{\mathbf{p}^2}{2m_e} + V(r), \tag{1.76}$$

where m_e is the mass of the electron. Since the mass of the electron is extremely small compared to that of the nucleus, the reduced mass is also assumed to be the mass of the electron, $\mu = m_e$. The potential V contains only the Coulomb interaction,

$$V = -\frac{e^2}{4\pi\epsilon_0 r}, \tag{1.77}$$

where e is the charge of the electron, ϵ_0 is the dielectric constant in a vacuum, and r is the distance from the nucleus to the electron. Similarly to Eq. (1.70), the eigenfunction of the electronic motion is provided as

$$\Psi_{\mathbf{p}}(\mathbf{r}) = Y_\lambda(\theta, \phi) R_{\mathbf{p}\lambda}(\mathbf{r}), \tag{1.78}$$

where the angular function Y_λ is the spherical harmonic function.

The difference between the electronic and molecular rotational motions is in the radial function $R_{p\lambda}$. In the molecular rotation motions, the radial function is assumed to be uninvolved in the quantum nature because of the typical deep potentials of interatomic bonds. However, this assumption is not appropriate for the electronic motions. Substituting Eq. (1.76) into Eq. (1.69) and (1.74) leads to the Schrödinger equation for the radial direction in polar coordinates (r, θ, ϕ),

$$\left(\frac{d^2}{dr^2} + \frac{2}{r}\frac{d}{dr}\right) R_{nl} + \frac{2m_e}{\hbar^2}\left[E + \frac{Ze^2}{r} - \frac{l(l+1)\hbar^2}{2m_e r^2}\right] R_{nl} = 0, \quad (1.79)$$

where the radial function $R_{p\lambda}$ is a complicated function with natural numbers n and l,

$$R_{nl}(r) = -\left[\left(\frac{2Z}{n}\right)^3 \frac{(n-l-1)!}{2n[(n+l)!]^3}\right]^{1/2} \exp\left(-\frac{Zr}{n}\right)\left(\frac{2Zr}{n}\right)^l L_{n+l}^{2l+1}\left(\frac{2Zr}{n}\right). \quad (1.80)$$

L_{n+l}^{2l+1} is Laguerre's adjoint polynomial,

$$L_{n+l}^{2l+1}(\rho) = \sum_{k=0}^{n-l-1} (-1)^{k+2l+1} \frac{[(n+l)!]^2 \rho^k}{(n-l-1-k)!(2l+1+k)!k!}. \quad (1.81)$$

It is significant to note that this polynomial is restricted to $n-l-1 \geq 0$, i.e. $n \geq l+1$. The energy eigenvalue corresponding to this eigenfunction is given as

$$E = -\frac{m_e c^2 (Z\alpha)^2}{2n^2}. \quad (1.82)$$

This is identical to the energy that Bohr determined by assuming the quantized periodic orbits of electron, although such an assumption is not required for the Schrödinger equation. Moreover, since the energy in Eq. (1.82) does not depend on the azimuthal quantum number l, this radial function is energetically degenerate; e.g., it is single ($l = 0$) for $n = 1$ but double ($l = 0, 1$) for $n = 2$. As mentioned above, the spherical harmonic function Y_{lm_l} is further degenerate for l in terms of the magnetic quantum number m_l ($-l \leq m \leq l$); e.g., it is single ($m_l = 1$) for $l = 0$ but triple ($m_l = -1, 0, 1$) for $l = 1$. In total, *the number of energetically degenerate radial functions for the azimuthal quantum number* l *is* $2l + 1$, *and the number of energetically degenerate total wavefunctions for the principal quantum number* n *is* n^2. Figure 1.6 illustrates the density distributions of the radial functions assigned to these quantum numbers. The figure clearly shows that the peaks of the density distributions are situated away from the nucleus as the number of n increases

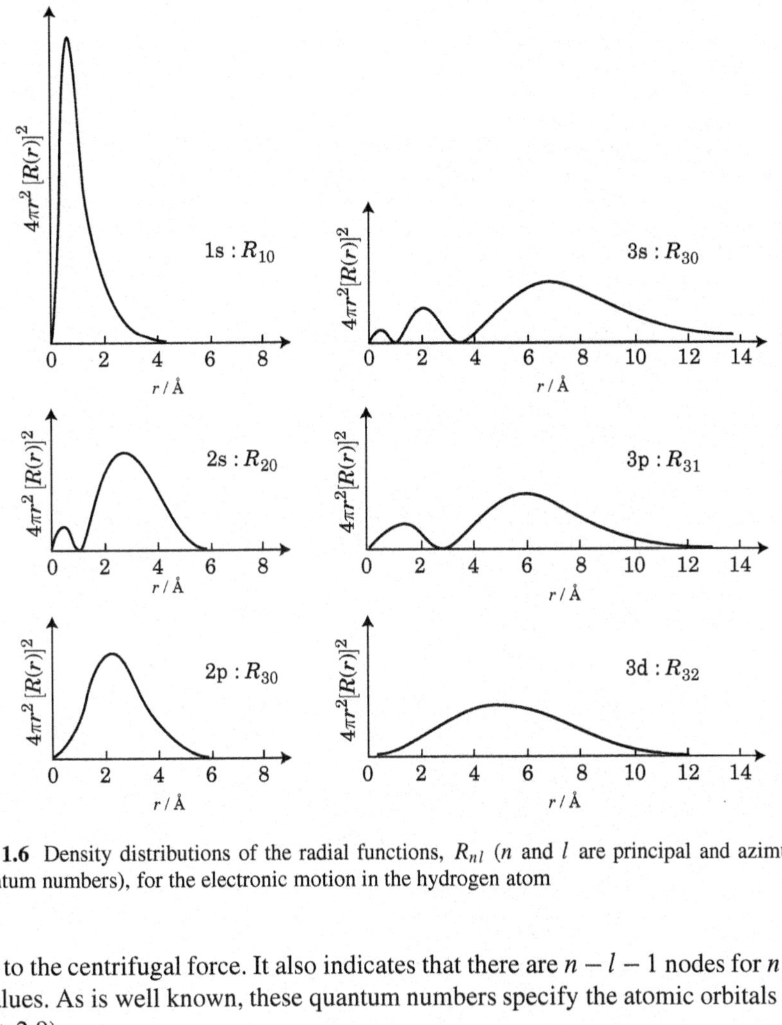

Fig. 1.6 Density distributions of the radial functions, R_{nl} (n and l are principal and azimuthal quantum numbers), for the electronic motion in the hydrogen atom

due to the centrifugal force. It also indicates that there are $n - l - 1$ nodes for n and l values. As is well known, these quantum numbers specify the atomic orbitals (see Sect. 2.9).

References

Alder, B.J., Wainwright, T.E.: J. Chem. Phys. **31**, 459–466 (1959)

Andersson, K., Malmqvist, P.A., Roos, B.O., Sadlej, A.J., Wolinski, K.: J. Phys. Chem. **94**, 5483–5488 (1990)

Asimov, I.: A Short History of Chemistry. Greenwood Press Reprint, Westport (1979)

Aspect, A., Grangier, P., Roger, G.: Phys. Rev. Lett. **49**, 91–94 (1982)

Becke, A.D.: Phys. Rev. A, **38**, 3098–3100 (1988)

Becke, A.D.: J. Chem. Phys. **98**, 5648–5652 (1993)

Becke, A.D.: J. Chem. Phys. **107**, 8554–8560 (1997)

Bell, J.: Physics **1**, 195–200 (1964)
Bell, R.P.: Proc. R. Soc. Lond. A **154**, 414–429 (1936)
Bohm, D.: Phys. Rev. **85**, 166–179 (1952)
Bohr, N.: Philos. Mag. **21**, 1–25 (1913)
Born, M., Heisenberg, W., Jordan, P.: Z. Phys. **35**, 557–615 (1926)
Boys, S.F.: Proc. R. Soc. Lond. A. **200**, 542–554 (1950)
Car, R., Parrinello, M.: Phys. Rev. Lett. **55**, 2471–2474 (1985)
Casida, M.E.: In: Seminario, J.J. (ed.) Recent Developments and Applications of Modern Density Functional Theory. Elsevier, Amsterdam (1996)
Čížek, J.: J. Chem. Phys. **45**, 4256–4266 (1966)
Compton, A.: Phys. Rev. **21**, 483–502 (1923)
Condon, E.U.: Phys. Rev. **36**, 1121–1133 (1930)
Coulson, C.A.: Proc. Camb. Philos. Soc. **34**, 204–212 (1938)
Davidson, E.R.: J. Comput. Phys. **17**, 87–94 (1975)
Davisson, C., Germer, L.H.: Phys. Rev. **30**, 705–740 (1927)
Dirac, P.A.M.: Proc. R. Soc. A **117**, 610–624 (1928)
Dirac, P.A.M.: Camb. Philos. Soc. **26**, 376–385 (1930)
Einstein, A.: Ann. Phys. **17**, 132–148 (1905)
Einstein, A., Podolsky, B., Rosen, N.: Phys. Rev. **47**, 777–780 (1935)
Ekeland, I.: Le Meilleur des Mondes Possibles (Japanese). Editions du Seuil (2009)
Evans, M.G., Polanyi, M.: Trans. Faraday Soc. **34**, 11–29 (1938)
Everett, H.: Rev. Mod. Phys. **29**, 454–462 (1957)
Eyring, H.: J. Chem. Phys. **3**, 107–115 (1935)
Fermi, E.: Z. Phys. **48**, 73–79 (1928)
Feynman, R.P.: Rev. Mod. Phys. **20**, 367–387 (1948)
Fock, V.: Z. Phys. **61**, 126–148 (1930)
Foster, J.M., and Boys, S.F.: Rev. Mod. Phys. **32**, 300–302 (1960)
Frenkel, J.: Wave Mechanics: Advanced General Theory. Clarendon Press, Oxford (1934)
Fukui, K., Yonezawa, T., Shingu, H.: J. Chem. Phys. **20**, 722–725 (1952)
Gasiorowicz, S.: Quantum Physics 2nd edn. Wiley, New York (1996)
Hall, G.G.: Proc. R. Soc. Lond. A **205**, 541–552 (1951)
Hartree, D.R.: Math. Proc. Camb. Philos. Soc. **24**, 89–132; 426–437 (1928)
Heine, V.: Solid State Phys. **24**, 1–36 (1970)
Heisenberg, W.: Z. Phys. **33**, 879–893 (1925)
Heisenberg, W.: Z. Phys. **38**, 411–426 (1926)
Heisenberg, W.: Z. Phys. **43**, 172–198 (1927)
Hirao, K.: Chem. Phys. Lett. **190**, 374–380 (1992)
Hoffmann, R.: J. Chem. Phys. **39**, 1397–1412 (1963)
Hohenberg, P., Kohn, W.: Phys. Rev. B **136**, 864–871 (1964)
Hückel, E.: Z. Phys. **60**, 423–456 (1930)
Hund, F.: Z. Phys. **36**, 657–674 (1926)
Iikura, H., Tsuneda, T., Yanai, T., Hirao, K.: J. Chem. Phys. **115**(8), 3540–3544 (2001)
Jammer, M.: The Philosophy of Quantum Mechanics: The Interpretations of Quantum Mechanics in Historical Perspective. Wiley, Chichester (1974)
Janak, J.F.: Phys. Rev. B **103**, 7165–7168 (1978)
Jönsson, C.: Z. Phys. **161**, 454–474 (1961)
Kahn, L.R., Goddard, W.A.: J. Chem. Phys. **56**, 2685–2701 (1972)
Kochen, S., Specker, E.P.: J. Math. Mech. **17**, 59–87 (1967)
Kohn, W., Sham, L.J.: Phys. Rev. A **140**, 1133–1138 (1965)
Koopmans, T.: Physica **1**, 104–113 (1934)
Lee, C., Yang, W., Parr, R.G.: Phys. Rev. B **37**, 785–789 (1988)
Lennard-Jones, J.E.: Trans. Faraday Soc. **25**, 668–676 (1929)
Levy, M.: Proc. Natl. Acad. Sci. USA **76**, 6062–6065 (1979)
Löwdin, P.-O.: Phys. Rev. **97**, 1509–1520 (1955)

Marcus, R.A.: J. Chem. Phys. **24**, 979–989 (1956)
Møller, C., Plesset, M.S.: Phys. Rev. **46**, 618–622 (1934)
Mulliken, R.S.: Phys. Rev. **29**, 637–649 (1927)
Mulliken, R.S.: J. Chem. Phys. **23**, 1833–1840 (1955)
Nakano, H.: J. Chem. Phys. **99**, 7983–7992 (1993)
Nakatsuji, H., Hirao, K.: J. Chem. Phys. **68**, 2053–2065 (1978)
Noether, E.A.: Nach. Ges. Wiss. Göttingen Math. Phys. **2**, 235–257 (1918)
Pariser, R., Parr, R.: J. Chem. Phys. **21**, 767–776 (1953)
Pauling, L.: Proc. Natl. Acad. Sci. USA **14**, 359–362 (1928)
Perdew, J.P., Zunger, A.: Phys. Rev. B **23**, 5048–5079 (1981)
Pople, J.A.: Trans. Faraday Soc. **49**, 1375–1385 (1953)
Pople, J.A., Nesbet, R.K.: J. Chem. Phys. **22**, 571–572 (1954)
Roothaan, C.C.J.: Rev. Mod. Phys. **23**, 69–89 (1951)
Rowe, D.J.: Rev. Mod. Phys. **40**, 153–166 (1968)
Runge, E., Gross, E.K.U.: Phys. Rev. Lett. **52**, 997–1000 (1984)
Schrödinger, E.: Phys. Rev. **28**, 1049–1070 (1926)
Schrödinger, E.: Naturwissenschaften **23**, 807–849 (1935)
Sinanoğlu, O.: Adv. Chem. Phys. **6**, 315–412 (1964)
Slater, J.C.: Phys. Rev. **32**, 339–348 (1928)
Slater, J.C.: Phys. Rev. **34**, 1293–1322 (1929)
Slater, J.C.: Phys. Rev. **35**, 210–211 (1930)
Thomas, L.H.: Proc. Camb. Philos. Soc. **23**, 542–548 (1927)
Vignale, G., Kohn, W.: Phys. Rev. Lett. **77**, 2037–2040 (1996)
von Neumann, J.L.: Mathematische Grundlagen der Quantenmechanik. Springer, Berlin (1957)
von Weizsäcker, C.F.: Z. Phys. **96**, 431–458 (1935)
Vosko, S.H., Wilk, L., Nusair, M.: Can. J. Phys. **58**, 1200–1211 (1980)
Warshell, A., Levitt, M.: J. Mol. Biol. **103**, 227–249 (1976)
Whitten, J.L., Hackmeyer, M.: J. Chem. Phys. **51**, 5584–5596 (1969)
Woodward, R.B., Hoffmann, R.: J. Am. Chem. Soc. **87**, 395–397 (1965)
Yang, W.: Phys. Rev. Lett. **66**, 1438–1441 (1991)

Chapter 2
Hartree–Fock Method

2.1 Hartree Method

For atoms other than the hydrogen atom, as treated in Sect. 1.9, we can obtain the quantum states of electronic motions in theory by constructing the Hamiltonian operators in a similar fashion and solving the Schrödinger equation. However, when we actually solve the Schrödinger equation for these atoms, we are faced with a serious problem: the *three-body problem*. That is, the state of motion cannot be solved analytically for systems in which three or more distinct masses interact. This three-body problem is not unique to quantum mechanics but is a classic problem in analytical physics. As described in Sect. 1.3, Poincaré proved that there are very few classical mechanics problems for which the equations of motions can be solved analytically. This is due to the three-body problem. Poincaré addressed this three-body problem in classical mechanics with the idea that only periodical solutions can be solved by the least (stationary)-action principle (Ekeland 2009). On the other hand, Hartree suggested an idea to solve the three-body problem for atoms containing multiple electrons.

In 1928, 2 years after the Schrödinger equation was published, Hartree proposed a method solving this equation for multiple-electron systems, based on fundamental physical principles: the *Hartree method* (Hartree 1928). Let us consider the electronic motion of a helium atom (Fig. 2.1). The Hamiltonian operator of this atom is given as

$$\hat{H} = -\frac{\nabla_1^2}{2} - \frac{\nabla_2^2}{2} + V_{\mathrm{ne}}(\mathbf{r}_1) + V_{\mathrm{ne}}(\mathbf{r}_2) + V_{\mathrm{ee}}(\mathbf{r}_1, \mathbf{r}_2), \tag{2.1}$$

where $\mathbf{r}_n$ is the position vector of the n-th electron and ∇_n is the gradient vector operator for the n-th electron. On the right-hand side of Eq. (2.1), the first two terms are kinetic energy operators, and the next two terms are nuclear-electron electrostatic interaction potentials. These are one-electron terms collectively called *one-electron operators*. On the other hand, the last term is an electron–electron

T. Tsuneda, *Density Functional Theory in Quantum Chemistry*,
DOI 10.1007/978-4-431-54825-6_2, © Springer Japan 2014

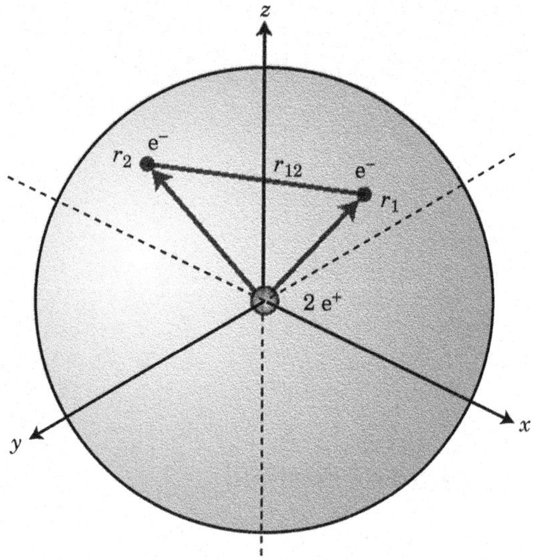

Fig. 2.1 Electrons in a helium atom, where $\mathbf{r}_i$ is the coordinate vector of the i-th electron, r_{12} is the distance between two electrons, and e^- and e^+ are negative and positive charges

electrostatic interaction potential called a *two-electron operator*. Notice that *atomic units* are used in this equation for the first time. Atomic units are a convenient system of units usually used in electronic property calculations. In this system, the *electron mass* m_e*, electron charge* e*, reduced Planck constant* $\hbar = h/2\pi$*, and Coulomb force constant* $1/(4\pi\epsilon_0)$ *are assumed to be 1.* Henceforth, let us consider the mechanics in atomic units for in order to reduce the complexity. In Eq. (2.1), potentials V_{ne} and V_{ee} are written as

$$V_{ne}(\mathbf{r}) = -\frac{2}{r} \tag{2.2}$$

and

$$V_{ee}(\mathbf{r}_{12}) = \frac{1}{r_{12}}, \tag{2.3}$$

where $r = |\mathbf{r}|$ and $r_{12} = |\mathbf{r}_2 - \mathbf{r}_1|$. That is, Hartree assumed that each electron moves in the averaged potential of the electrostatic interactions with surrounding electrons and suggested the *independent electron approximation* which approximates the averaged potential as an effective potential V_{eff}. For this approximation, the Hamiltonian operator is divided into terms for the different electrons, as follows:

$$\begin{aligned}\hat{H} &= \left[-\frac{\nabla_1^2}{2} + V_{ne}(r_1) + V_{eff}(\mathbf{r}_1)\right] + \left[-\frac{\nabla_2^2}{2} + V_{ne}(r_2) + V_{eff}(\mathbf{r}_2)\right] \\ &= \hat{h}(\mathbf{r}_1) + \hat{h}(\mathbf{r}_2),\end{aligned} \tag{2.4}$$

where $h(\mathbf{r}_i)$ is the Hamiltonian operator for the i-th electron. In this Hamiltonian operator, the wavefunction is represented as the product of different electronic motion wavefunctions,

$$\Psi(\mathbf{r}_1, \mathbf{r}_2) = \phi_1(\mathbf{r}_1)\phi_2(\mathbf{r}_2). \tag{2.5}$$

In this case, the expectation value of the Hamiltonian operator is given as

$$\begin{aligned} E &= \frac{\int d^3\mathbf{r}_1 d^3\mathbf{r}_2 \Psi^*(\mathbf{r}_1, \mathbf{r}_2)\hat{H}\Psi(\mathbf{r}_1, \mathbf{r}_2)}{\int d^3\mathbf{r}_1 d^3\mathbf{r}_2 \Psi^*(\mathbf{r}_1, \mathbf{r}_2)\Psi(\mathbf{r}_1, \mathbf{r}_2)} \\ &= \frac{\sum_{i=1}^{2} \int d^3\mathbf{r}_i \phi_i^*(\mathbf{r}_i)\hat{h}(\mathbf{r}_i)\phi_i(\mathbf{r}_i)}{\sum_{i=1}^{2} \int d^3\mathbf{r}_i \phi_i^*(\mathbf{r}_i)\phi_i(\mathbf{r}_i)}. \end{aligned} \tag{2.6}$$

With the variational method, the Schrödinger equation for obtaining the set of one-electron wavefunctions $\{\phi_i\}$, which makes this expectation value stationary, is represented as

$$\hat{h}(\mathbf{r}_i)\phi_i(\mathbf{r}_i) = \epsilon_i \phi_i(\mathbf{r}_i). \tag{2.7}$$

Since the eigenvalue of Eq. (2.7), ϵ_i, is interpreted as the eigenenergy for the motion of the i-th electron, the total wavefunction can be obtained by solving the eigenequation for each electron. This one-electron wavefunction ϕ_i and the corresponding eigenenergy ϵ_i are later called the *orbital* and *orbital energy*, respectively. This equation also indicates that the total eigenenergy is the sum of the orbital energies corresponding to different electronic motions,

$$\hat{H}\Psi = (\epsilon_1 + \epsilon_2)\phi_1\phi_2 = \epsilon\Psi. \tag{2.8}$$

This theory is called the Hartree method.

Although the effective potential is replaced with a virtual potential in Eq. (2.4), this does not necessarily require such a replacement. The effective potential corresponding to the wavefunction in Eq. (2.5) is derived as

$$V_{\text{eff}}(\mathbf{r}_i) = \sum_j \int d^3\mathbf{r}_j \frac{|\phi_j(\mathbf{r}_j)|^2}{|\mathbf{r}_i - \mathbf{r}_j|}. \tag{2.9}$$

The above discussion is applicable to the theory using this potential, because the motions of electrons are independent for this potential. In general, a theory using a potential derived with no semiempirical parameters is called an *ab initio theory* ("ab initio" means "from the beginning" in Latin). Solving this ab initio Hartree equation is not as straightforward as it might appear at first glance, because the effective potential in Eq. (2.9) is represented by the wavefunctions of other electronic motions. That is, Eq. (2.7) is a nonlinear equation in which the operator depends on the solution of this equation. Hartree incorporated the *self-consistent field (SCF) method* to solve this equation. For the SCF method, further details are described in Sect. 2.4. This ab initio Hartree method reproduces the shapes of atomic orbitals reasonably close to those of state-of-the-art theories that take into account electron correlation, as mentioned later. However, it was found that the energies given by this method are too inaccurate to be used in the analyses of chemical reactions and properties. It becomes clear that the inaccurate energies are mainly attributed to the neglect of electron–electron exchange interactions (see Sect. 2.3).

2.2 Molecular Orbital Theory

What motional states do electrons have in molecules? Let us consider the case of the hydrogen molecule (Fig. 2.2). With the Hartree method, the Hamiltonian operator for the hydrogen molecule is given by

$$\hat{H} = \hat{h}(\mathbf{r}_1) + \hat{h}(\mathbf{r}_2) + V_{\mathrm{nn}}(R_{\mathrm{AB}}), \tag{2.10}$$

where R_{AB} is the distance between two atomic nuclei, A and B. Neglecting nuclear motions is one of the most important and convenient approximations in considering the electronic motions in molecules. Since even the lightest atomic nucleus, hydrogen, is 1836 times heavier than the electron, electrons are considered to move overwhelmingly faster than nuclei do. Actually, the kinetic energy of the nucleus has already been neglected when electronic motions are considered in Eq. (2.1). The approximation neglecting atomic nuclear motions to consider electronic motions is called the *adiabatic approximation*, based on the interpretation that the absence

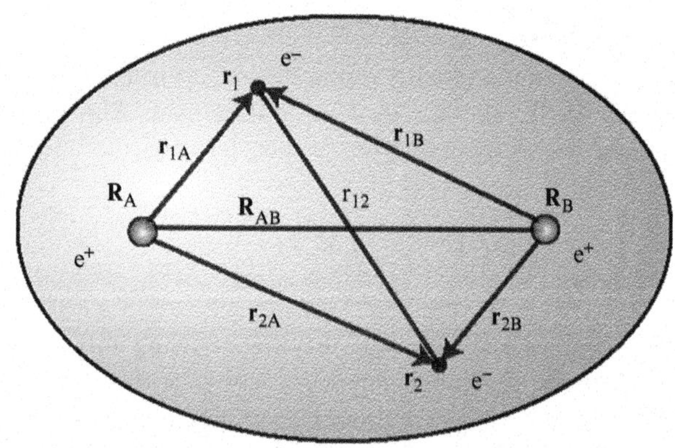

Fig. 2.2 Electrons in hydrogen molecule, where $\mathbf{R}_I$ is the coordinate vector of the I-th nuclear, r_{iI} is the distance between the i-th electron and I-th nuclear, and R_{AB} is the distance between nuclei A and B

of atomic nuclear motion means that there is no heat energy interchange with the outside environment, or also as the *Born–Oppenheimer approximation*, capping the names of the developers (Born and Oppenheimer 1927). This approximation is so efficient that it hardly affects the chemical properties of electronic ground states. Therefore, this approximation is used by default in quantum chemical calculations. With this adiabatic approximation, since the nuclear–nuclear interactions V_{nn} can be neglected, Eq. (2.10) becomes

$$\hat{H} = \hat{h}(\mathbf{r}_1) + \hat{h}(\mathbf{r}_2), \tag{2.11}$$

which appears similar to the Hamiltonian of the helium atom. However, the equation is more difficult to solve than it appears, because the hydrogen molecule has two atomic nuclei. Actually, the position vectors of the electrons must be considered for each atomic nucleus, although these are represented as $\mathbf{r}_n$ in Eq. (2.11).

To solve this problem, the concept of the *molecular orbital* was introduced. The concept of the molecular orbital was used by Hund for the first time to explain the electronic states of molecules (Hund 1926). Mulliken summarized this concept in English, which helped to hasten its introduction to the world at large (Mulliken 1927). He is also the one who coined the term "orbital" in 1932 to indicate its similarity to an orbit. In 1929, Lennard–Jones suggested an ansatz (educated guess) that molecular eigenfunctions are representable as the linear combination of atomic eigenfunctions, which became the basis of the linear combination of atomic orbitals–molecular orbital (*LCAO–MO*) *approximation* for describing molecular orbitals in terms of the atomic orbitals (Lennard-Jones 1929). Coulson reported the variational calculation results of the hydrogen molecule using the LCAO–MO approximation for the first time (Coulson 1938). This was the first usage of *molecular orbital theory*.

Based on the LCAO–MO approximation, the molecular orbital of the hydrogen molecule is written as

$$\phi = C_1\chi_1 + C_2\chi_2, \tag{2.12}$$

as shown in Fig. 2.3, where χ_I is the atomic orbital centered on the I-th atomic nucleus, and C_I is the corresponding molecular orbital coefficient. In Eq. (2.12), the atomic orbitals and molecular orbital coefficients are assumed to be real. Substituting this molecular orbital ϕ into Eq. (2.6) gives

$$\epsilon = \frac{C_1^2 h_{11} + C_2^2 h_{22} + 2C_1C_2h_{12}}{C_1^2 + C_2^2 + 2C_1C_2S_{12}}, \tag{2.13}$$

where h_{ij} and S_{12} are given by

$$h_{ij} = \int d^3\mathbf{r}\chi_i\hat{h}\chi_j < 0 \tag{2.14}$$

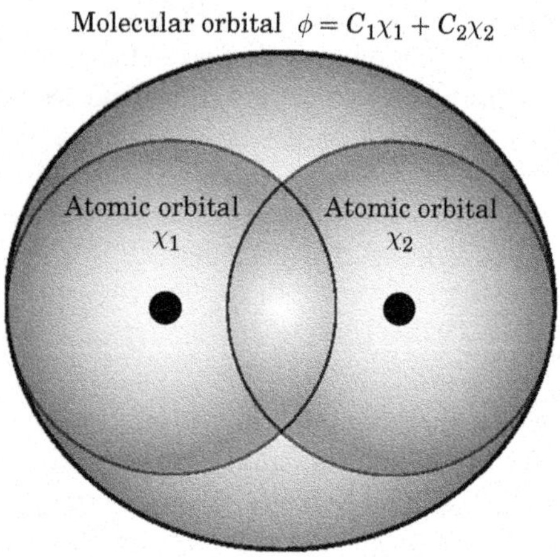

Fig. 2.3 Molecular orbital of the hydrogen molecule, ϕ, composed of the atomic orbitals of hydrogen atoms, χ_1 and χ_2, for the LCAO–MO approximation

and

$$S_{12} = \int d^3\mathbf{r}\chi_1\chi_2 > 0, \tag{2.15}$$

and ϵ indicates the molecular orbital energy. Since the coefficients C_1 and C_2 minimizing this ϵ are determined by

$$\partial\epsilon/\partial C_1 = \partial\epsilon/\partial C_2 = 0, \tag{2.16}$$

Eq. (2.13) gives

$$(h_{11} - \epsilon)C_1 + (h_{12} - \epsilon S_{12})C_2 = 0, \tag{2.17}$$

and

$$(h_{12} - \epsilon S_{12})C_1 + (h_{22} - \epsilon)C_2 = 0. \tag{2.18}$$

The solutions for these equations, except for the case of $C_1 = C_2 = 0$, are obtained by

$$\begin{vmatrix} h_{11} - \epsilon & h_{12} - \epsilon S_{12} \\ h_{12} - \epsilon S_{12} & h_{22} - \epsilon \end{vmatrix} = 0. \tag{2.19}$$

By solving this equation, the orbital energies of the hydrogen molecule are given by

$$\epsilon_{\pm} = \frac{h_{11} \pm h_{12}}{1 \pm S_{12}} \tag{2.20}$$

and the corresponding normalized molecular orbitals are

$$\phi_{\pm} = \frac{\chi_1 \pm \chi_2}{\sqrt{2 \pm 2S_{12}}}. \tag{2.21}$$

For the orbital energies, ϵ_+ is supposed to have a lower energy, because S_{12} and h_{12} are generally given as $0 < S_{12} \ll 1$ and $h_{12} \approx S_{12}(h_{11} + h_{22})/2 < 0$. The electron occupancy of the molecular orbitals obeys the *Pauli exclusion principle* (Pauli 1925). This principle indicates that the electron has *spin* as the fourth quantum number, and no two electrons can have the identical set of quantum numbers simultaneously. For a hydrogen molecule containing only two electrons, opposite-spin electrons occupy the molecular orbital ϕ_+. Therefore, the wavefunction for the hydrogen molecule is given by the

$$\Phi = \phi_+(\mathbf{r}_1)\phi_+(\mathbf{r}_2) \tag{2.22}$$

$$= \frac{1}{2 + 2S_{12}} \{\chi_1(\mathbf{r}_1) + \chi_2(\mathbf{r}_1)\} \{\chi_1(\mathbf{r}_2) + \chi_2(\mathbf{r}_2)\}. \tag{2.23}$$

The total electronic energy of the hydrogen molecule is calculated as

$$E = 2\epsilon_+ = \frac{2(h_{11} + h_{12})}{1 + S_{12}}. \tag{2.24}$$

Notice that this is based not on the ab initio method but on the original Hartree method. With the ab initio method, the double-counted electron–electron Coulomb interaction must be removed from the total energy. In this way, the electronic states of molecules can be calculated using the LCAO–MO approximation.

As described above, the binding energy of the hydrogen molecule equals the orbital energy. However, *the analysis of binding energies based on orbital energies is applicable only to the limited case in which only two electrons are involved in the bond, like the hydrogen molecule*. Discussing other types of binding energies requires consideration of the energy contributions from other orbitals. For instance, it is generally difficult to evaluate molecular bond strength from the energetic stabilities of molecular orbital energies before and after reactions. In most cases, there is no clear relation between orbital and binding energies. However, this does not negate the value of the conceptual reaction analysis using molecular orbital images. *Since molecular orbitals have a relatively small dependence of the shapes on the type of calculation method, these are assumed to be correct in most cases, if the calculated system is not so large*. It is therefore possible to obtain information on reaction sites and reactivities of molecules from electron distributions given by molecular orbitals, as seen in the analysis on the Frontier orbital theory (Fukui et al. 1952). Note, however, that since orbital energies significantly depend on the calculation method, as described in Sect. 7, the order of the orbital energy levels are easily changed, especially if there are virtual orbitals.

2.3 Slater Determinant

In 1926, Heisenberg and Dirac independently proposed that *the wavefunction of electronic motions must be antisymmetric* (the sign of the wavefunction becomes opposite for the exchange of electrons) *to satisfy the Pauli exclusion principle naturally and therefore should be represented as a determinant* (Heisenberg 1926; Dirac 1926). The new electron–electron interaction resulting from the antisymmetrization is called the *exchange interaction*. Slater constructed a general method for solving the Schrödinger equation based on the normalized determinant representing the antisymmetrized wavefunction (Slater 1929). In this study, the *exchange integral* was also suggested for the first time.

Let us consider the exchange of electrons by taking the helium atom as an example. The Hamiltonian operator in Eq. (2.1) does not change by exchanging the coordinates of two electrons:

$$\hat{H}(\mathbf{r}_1, \mathbf{r}_2) = \hat{H}(\mathbf{r}_2, \mathbf{r}_1). \tag{2.25}$$

Since this leads to

$$\hat{H}(\mathbf{r}_1, \mathbf{r}_2)\Psi(\mathbf{r}_2, \mathbf{r}_1) = E\Psi(\mathbf{r}_2, \mathbf{r}_1), \tag{2.26}$$

it is easily proven that the Hamiltonian operator is commutative with the exchange operator replacing the positions of two electrons $\hat{P}_{12}$, as follows:

$$\hat{H}(\mathbf{r}_1, \mathbf{r}_2)\hat{P}_{12}\Psi(\mathbf{r}_1, \mathbf{r}_2) = \hat{P}_{12}\hat{H}(\mathbf{r}_1, \mathbf{r}_2)\Psi(\mathbf{r}_1, \mathbf{r}_2), \tag{2.27}$$

i.e.,

$$\left[\hat{H}, \hat{P}_{12}\right]\Psi = \left(\hat{H}\hat{P}_{12} - \hat{P}_{12}\hat{H}\right)\Psi = 0. \tag{2.28}$$

This indicates that the Hamiltonian and exchange operators have simultaneous eigenstates. Moreover, $\hat{P}_{12}^2 = 1$ is found by considering that the positions of two electrons are reverted by two exchanges. This leads to the eigenvalues of $\hat{P}_{12}$ being ± 1. Therefore, there are two different wavefunctions for the exchange operator $\hat{P}_{12}$: the symmetric wavefunction corresponding to the $+1$ eigenvalue,

$$\Psi^{(S)}(\mathbf{r}_1, \mathbf{r}_2) = \frac{1}{\sqrt{2}}\left\{\Psi(\mathbf{r}_1, \mathbf{r}_2) + \Psi(\mathbf{r}_2, \mathbf{r}_1)\right\}, \tag{2.29}$$

and the antisymmetric wavefunction corresponding to the -1 eigenvalue,

$$\Psi^{(A)}(\mathbf{r}_1, \mathbf{r}_2) = \frac{1}{\sqrt{2}}\left\{\Psi(\mathbf{r}_1, \mathbf{r}_2) - \Psi(\mathbf{r}_2, \mathbf{r}_1)\right\} = -\Psi^{(A)}(\mathbf{r}_2, \mathbf{r}_1), \tag{2.30}$$

where $1/\sqrt{2}$ is the normalization constant.

Next, let us consider the Hartree wavefunction in Eq. (2.5). The symmetric and antisymmetric wavefunctions are given by

$$\Psi^{(S)}(\mathbf{r}_1, \mathbf{r}_2) = \frac{1}{\sqrt{2}} \{\phi_1(\mathbf{r}_1)\phi_2(\mathbf{r}_2) + \phi_1(\mathbf{r}_2)\phi_2(\mathbf{r}_1)\} \tag{2.31}$$

and

$$\Psi^{(A)}(\mathbf{r}_1, \mathbf{r}_2) = \frac{1}{\sqrt{2}} \{\phi_1(\mathbf{r}_1)\phi_2(\mathbf{r}_2) - \phi_1(\mathbf{r}_2)\phi_2(\mathbf{r}_1)\}, \tag{2.32}$$

respectively. In these wavefunctions, only the antisymmetric wavefunction satisfies the Pauli exclusion principle. This is because wavefunctions must be zero for the case in which the same electron occupies the same orbital, i.e. $\mathbf{r}_1 = \mathbf{r}_2$. As a result, *electronic motions always have antisymmetric wavefunctions.* The antisymmetric wavefunction in Eq. (2.32) can be written as a determinant,

$$\Psi(\mathbf{r}_1, \mathbf{r}_2) = \frac{1}{\sqrt{2}} \begin{vmatrix} \phi_1(\mathbf{r}_1) & \phi_1(\mathbf{r}_2) \\ \phi_2(\mathbf{r}_1) & \phi_2(\mathbf{r}_2) \end{vmatrix}. \tag{2.33}$$

For the case of three or greater electrons, the antisymmetric wavefunction can be described in a determinant form,

$$\Psi(\mathbf{r}_1, \mathbf{r}_2, \ldots, \mathbf{r}_N) = \frac{1}{\sqrt{N!}} \begin{vmatrix} \phi_1(\mathbf{r}_1) & \phi_1(\mathbf{r}_2) & \ldots & \phi_1(\mathbf{r}_N) \\ \phi_2(\mathbf{r}_1) & \phi_2(\mathbf{r}_2) & \ldots & \phi_2(\mathbf{r}_N) \\ \vdots & \vdots & \ddots & \vdots \\ \phi_N(\mathbf{r}_1) & \phi_N(\mathbf{r}_2) & \ldots & \phi_N(\mathbf{r}_N) \end{vmatrix} \tag{2.34}$$

$$= \frac{1}{\sqrt{N!}} \det |\phi_1(\mathbf{r}_1)\phi_2(\mathbf{r}_2)\cdots\phi_N(\mathbf{r}_N)|. \tag{2.35}$$

This determinant is called the *Slater determinant* (Slater 1929).

2.4 Hartree–Fock Method

In 1930, Fock applied the Slater determinant to the Hartree method and proposed the *Hartree–Fock method*, which is one of the core theories in quantum chemistry (Fock 1930). Slater also suggested the same method independently in the same year (Slater 1930).

For simplicity, let us consider closed-shell systems, in which n orbitals are occupied by electrons two by two. The Slater determinants for closed-shell systems are given by defining orbitals as functions of the spatial and spin coordinates, as follows:

$$\Phi(\{\mathbf{r},\sigma\}) = \frac{1}{\sqrt{(2n)!}} \det|\phi_1(\mathbf{r}_1,\alpha)\phi_1(\mathbf{r}_2,\beta)\cdots\phi_n(\mathbf{r}_{2n-1},\alpha)\phi_n(\mathbf{r}_{2n},\beta)|, \quad (2.36)$$

where $\{\mathbf{r},\sigma\} = \{\mathbf{r}_1\sigma_1,\ldots,\mathbf{r}_{2n}\sigma_{2n}\}$. The expectation value of the corresponding Hamiltonian operator is written instead of Eq. (2.6) as

$$E = \frac{1}{(2n)!}\int d^3\{\mathbf{r}\}d\{\sigma\}\det|\phi_1^*(\mathbf{r}_1,\alpha)\cdots\phi_n^*(\mathbf{r}_{2n},\beta)|\,\hat{H} \times \det|\phi_1(\mathbf{r}_1,\alpha)\cdots\phi_n(\mathbf{r}_{2n},\beta)|, \quad (2.37)$$

where $d^3\{\mathbf{r}\} = d^3\mathbf{r}_1\cdots d^3\mathbf{r}_{2n}$ and $d\{\sigma\} = d\sigma_1\cdots d\sigma_{2n}$. Since electrons are indistinguishable, the integrals for the one-electron operator (*one-electron integrals*) are $2n$-tuplicate, and the integrals for the two-electron operator (*two-electron integrals*) are $\{2n(2n-1)/2\}$-tuplicate. Therefore, the expectation value is transformed to

$$E = \frac{1}{(2n)!}\int d^3\{\mathbf{r}\}d\{\sigma\}\det|\phi_1^*(\mathbf{r}_1,\alpha)\cdots\phi_n^*(\mathbf{r}_{2n},\beta)|\,\hat{H} \times \det|\phi_1(\mathbf{r}_1,\alpha)\cdots\phi_n(\mathbf{r}_{2n},\beta)| \quad (2.38)$$

$$= \frac{1}{(2n-1)!}\int d^3\{\mathbf{r}\}d\{\sigma\}\det|\phi_1^*(\mathbf{r}_1,\alpha)\cdots\phi_n^*(\mathbf{r}_{2n},\beta)| \times\left(-\frac{\nabla^2}{2}+V_{\mathrm{ne}}\right)\det|\phi_1(\mathbf{r}_1,\alpha)\cdots\phi_n(\mathbf{r}_{2n},\beta)|$$
$$+\frac{1}{2(2n-2)!}\int d^3\{\mathbf{r}\}d\{\sigma\}\det|\phi_1^*(\mathbf{r}_1,\alpha)\cdots\phi_n^*(\mathbf{r}_{2n},\beta)|\frac{1}{r_{12}} \times\det|\phi_1(\mathbf{r}_1,\alpha)\cdots\phi_n(\mathbf{r}_{2n},\beta)|. \quad (2.39)$$

The one-electron integrals in the first term of the right-hand side of Eq. (2.39) are $(2n-1)!$-tuplicate under the following conditions:

$$\int d^3\mathbf{r}_\lambda d\sigma\phi_k^*(\mathbf{r}_\lambda,\sigma)\phi_l(\mathbf{r}_\lambda,\sigma) = \delta_{kl} \quad (2.40)$$

and

$$\int d^3\mathbf{r}_\lambda d\sigma\phi_k^*(\mathbf{r}_\lambda,\sigma)\phi_l(\mathbf{r}_\lambda,\sigma'\neq\sigma) = 0. \quad (2.41)$$

Since the same spatial orbitals are used for α and β spins, the one-electron integrals are described without concern for spins as

$$
\begin{aligned}
(\text{One-electron integrals}) &= 2\sum_i^n \int d^3\mathbf{r}\phi_i^*(\mathbf{r})\left\{-\frac{1}{2}\nabla^2 + V_{\text{ne}}(\mathbf{r})\right\}\phi_i(\mathbf{r}) \\
&= 2\sum_i^n h_i.
\end{aligned}
\tag{2.42}
$$

Note that ϕ_k is switched from a spin orbital $\phi_k(\mathbf{r}, \sigma)$ to a spatial orbital $\phi_k(\mathbf{r})$. For the two-electron integrals of the second term in Eq. (2.39), it is not necessary to take spin into consideration for the combinations of different spin terms, which equal zero under the same conditions, similarly to the one-electron integrals. On the other hand, for the spatial terms, the combinations of exchanging two electrons remain besides those of the same spatial terms. These combinations are $(n-2)!$-tuplicate. The two-electron integrals are summarized as

$$
(\text{Two-electron integrals}) = \sum_{i,j}^n \left(2J_{ij} - K_{ij}\right), \tag{2.43}
$$

where J_{ij} and K_{ij} are

$$
J_{ij} = \int d^3\mathbf{r}_1 d^3\mathbf{r}_2 \phi_i^*(\mathbf{r}_1)\phi_j^*(\mathbf{r}_2)\frac{1}{r_{12}}\phi_i(\mathbf{r}_1)\phi_j(\mathbf{r}_2) = \langle ij|ij\rangle \tag{2.44}
$$

and

$$
K_{ij} = \int d^3\mathbf{r}_1 d^3\mathbf{r}_2 \phi_i^*(\mathbf{r}_1)\phi_j^*(\mathbf{r}_2)\frac{1}{r_{12}}\phi_j(\mathbf{r}_1)\phi_i(\mathbf{r}_2) = \langle ij|ji\rangle. \tag{2.45}
$$

Attaching the stationary condition for the minimal variation of orbital ϕ_i ($\phi_i \rightarrow \phi_i + \delta\phi_i$), $\delta E/\delta\phi_i = 0$, to the expectation value of the Hamiltonian operator containing one- and two-electron integrals, we obtain

$$
\left\{h_i + \sum_j^n \left(2\hat{J}_j - \hat{K}_j\right)\right\}\phi_i = \sum_j \phi_j \epsilon_{ji} \tag{2.46}
$$

$$
\left\{h_i^* + \sum_j^n \left(2\hat{J}_j^* - \hat{K}_j^*\right)\right\}\phi_i^* = \sum_j \phi_j^* \epsilon_{ji}^* \tag{2.47}
$$

from the variational principle. In Eqs. (2.46) and (2.47), $\hat{J}_j$ and $\hat{K}_j$ are called *Coulomb operator* and *exchange operator*, respectively, which are defined as

$$
\hat{J}_j(\mathbf{r}_1)\phi_i(\mathbf{r}_1) = \int d^3\mathbf{r}_2 \phi_j^*(\mathbf{r}_2)\phi_j(\mathbf{r}_2)\frac{1}{r_{12}}\phi_i(\mathbf{r}_1) \tag{2.48}
$$

and

$$\hat{K}_j(\mathbf{r}_1)\phi_i(\mathbf{r}_1) = \int d^3\mathbf{r}_2 \phi_j^*(\mathbf{r}_2)\phi_i(\mathbf{r}_2)\frac{1}{r_{12}}\phi_j(\mathbf{r}_1). \tag{2.49}$$

Note that the meaningful Hamiltonian operator is a Hermitian operator, giving real eigenenergies,

$$\epsilon_{ji} = \epsilon_{ij}^*. \tag{2.50}$$

This indicates that Eq. (2.47) is the complex conjugate of Eq. (2.46). Provided that the Fock matrix is defined as

$$\hat{F} = \hat{h} + \sum_j^n \left(2\hat{J}_j - \hat{K}_j\right), \tag{2.51}$$

the one-electron equation is derived as

$$\hat{F}\phi_i = \epsilon_i\phi_i. \tag{2.52}$$

Equation (2.52) is called the *Hartree–Fock equation*, and the operator $\hat{F}$ is called the *Fock operator*.

Using Eq. (2.52), the orbital energy ϵ_i is represented as

$$\epsilon_i = \int d^3\mathbf{r}_1 \phi_i^*(\mathbf{r}_1)\hat{F}\phi_i(\mathbf{r}_1) \tag{2.53}$$

$$= h_i + \sum_j^n \left(2J_{ij} - K_{ij}\right). \tag{2.54}$$

By this orbital energy, the total electron energy is written as

$$E = 2\sum_i^n \epsilon_i - \sum_{i,j}^n \left(2J_{ij} - K_{ij}\right) \tag{2.55}$$

$$= \sum_i^n \left(\epsilon_i + h_i\right). \tag{2.56}$$

Note that the total electron energy is not identical to the sum of the two orbital energies, because the latter doubly counts the electron–electron interactions. The total energy is calculated by adding the nucleus–nucleus repulsion energies to the total electron energy,

$$E_{\text{total}} = E + \sum_{A \neq B} \frac{Z_A Z_B}{R_{AB}}. \tag{2.57}$$

Since the Hartree–Fock equation is a nonlinear equation, similarly to the Hartree equation, it is usually solved by the SCF method. The Hartree–Fock SCF method is carried out via the following process:

1. Set up the information on the calculated molecular system (nuclear coordinates, nuclear charges, and number of electrons) and initial reference molecular orbitals $\{\phi_i^0\}$.
2. Calculate the two-electron operators of the Fock operator in Eq. (2.51) using the reference orbitals.
3. Solve Eq. (2.52) with the calculated Fock operator.
4. Compare the given molecular orbitals with the reference orbitals to determine if these orbitals are the eigenfunctions of this equation. The given orbitals are taken as the eigenfunctions if the differences of all the given orbitals to the corresponding reference orbitals are less than a given threshold value. If the orbitals have differences greater than the threshold, return to step 2 with the given orbitals as initial orbitals.

In general, the Hartree–Fock method indicates this SCF-based method. Despite the simplicity of the procedure, it soon became clear that solving this equation is not-trivial for usual molecular electronic systems. The Hartree–Fock equation essentially cannot be solved for molecules without computers. Actually, solving the Hartree–Fock equation for molecules had to await the appearance of general-purpose computers.

2.5 Roothaan Method

The preparation for computer-based science began following the practical use of computers represented by the appearance of commercial computers in 1950. In the following year (1951), Roothaan, an IBM researcher, developed a method for solving the Hartree–Fock equation on a computer, which is called the *Roothaan method* (Roothaan 1951). Since Hall independently suggested this method in the same year (Hall 1951), this is also called Roothaan–Hall method. This method is based on the character of von Neumann-type computer architecture, which efficiently performs matrix operations. A year earlier (1950), the *Gaussian-type basis function* was suggested by Boys (Boys 1950). Using the contracted nature of the Gaussian basis functions (see Sect. 2.6), the Roothaan method dramatically reduced the computational time of the Hartree–Fock calculations to realize the electronic state calculations of general molecules. After the development of this method, quantum chemistry began to make progress that tracked the growth of computer power.

Based on the LCAO–MO approximation given in Sect. 2.2, molecular orbitals $\{\phi_i\}$ can be expanded as

$$\phi_i(\mathbf{r}) = \sum_{p=1}^{n_{\mathrm{AO}}} \chi_p(\mathbf{r}) C_{pi}, \tag{2.58}$$

where n_{AO} is the number of atomic orbitals. The expansion coefficient C_{pi} is called the *molecular orbital (MO) coefficient*. Although $\{\chi_p\}$ is essentially the set of atomic orbitals, it is more efficient and general to use basis functions modeling atomic orbitals (For basis sets, see Sect. 2.6). Using the expansion of molecular orbitals, the Hartree–Fock equation is transformed to a matrix equation,

$$\mathbf{F}\mathbf{C}_i = \epsilon_i \mathbf{S}\mathbf{C}_i, \tag{2.59}$$

where the elements of matrix $\mathbf{F}$ and $\mathbf{S}$ are given as

$$\begin{aligned} F_{pq} &= \int d^3\mathbf{r} \chi_p^*(\mathbf{r}) \hat{F} \chi_q(\mathbf{r}) \\ &= \int d^3\mathbf{r} \chi_p^*(\mathbf{r}) \hat{h} \chi_q(\mathbf{r}) + \sum_{j=1}^{n} \int d^3\mathbf{r} \chi_p^*(\mathbf{r}) \left(2\hat{J}_j - \hat{K}_j \right) \chi_q(\mathbf{r}) \\ &= h_{pq} + \sum_{r,s=1}^{n_{\mathrm{basis}}} P_{sr} \left(\langle pr|qs \rangle - \frac{1}{2} \langle pr|sq \rangle \right), \end{aligned} \tag{2.60}$$

$$S_{pq} = \int d^3\mathbf{r} \chi_p^*(\mathbf{r}) \chi_q(\mathbf{r}). \tag{2.61}$$

In these equations, n_{basis} is the number of basis functions, and the integrals h_{pq}, $\langle pr|qs \rangle$ and $\langle pr|sq \rangle$ indicate

$$h_{pq} = \int d^3\mathbf{r} \chi_p^*(\mathbf{r}) \left\{ -\frac{1}{2} \nabla_r^2 + V_{\mathrm{ne}}(\mathbf{r}) \right\} \chi_q(\mathbf{r}), \tag{2.62}$$

$$\langle pr|qs \rangle = \int d^3\mathbf{r}_1 d^3\mathbf{r}_2 \chi_p^*(\mathbf{r}_1) \chi_r^*(\mathbf{r}_2) \frac{1}{r_{12}} \chi_q(\mathbf{r}_1) \chi_s(\mathbf{r}_2), \tag{2.63}$$

$$\langle pr|sq \rangle = \int d^3\mathbf{r}_1 d^3\mathbf{r}_2 \chi_p^*(\mathbf{r}_1) \chi_r^*(\mathbf{r}_2) \frac{1}{r_{12}} \chi_s(\mathbf{r}_1) \chi_q(\mathbf{r}_2). \tag{2.64}$$

P_{pq} is called the *density matrix*, which is defined using molecular orbital coefficients as

$$P_{pq} = 2 \sum_{j=1}^{n} C_{pj} C_{qj}^*. \tag{2.65}$$

The density matrix was first introduced by von Neumann (von Neumann 1927) and Landau (Landau 1927) independently to describe the quantum mechanical natures of statistical systems. The diagonal term of the density matrix in Eq. (2.65), P_{pp}, indicates the existence probability of an electron in the p-th molecular orbital.

For Eq. (2.59), the condition

$$|\mathbf{F} - \epsilon_i \mathbf{S}| = 0 \tag{2.66}$$

is required to make the coefficient matrix $\mathbf{C}_i$ give nonzero solutions. To solve Eq. (2.66), a transformation is usually used for Eq. (2.59). That is, the Fock matrix $\mathbf{F}$ is transformed to give matrix $\mathbf{F}'$ satisfying

$$|\mathbf{F}' - \epsilon_i \mathbf{E}| = 0, \tag{2.67}$$

where $\mathbf{E}$ is the unit matrix. By this transformation, Eq. (2.66) is easily solved by simply diagonalizing matrix $\mathbf{F}'$. The simplest way to give Eq. (2.67) is orthonormalizing basis sets beforehand. Provided that basis sets giving the unit matrix $\mathbf{E}$ for the overlap matrix $\mathbf{S}$ would be used, Eq. (2.66) is easily solved only by diagonalizing the Fock matrix. Such a basis set transformation is actually employed in many quantum chemistry calculation programs.

For using orthonormalized basis sets, the Roothaan equation is solved by the SCF method as follows:

1. Set up the information for the calculated molecular system (nuclear coordinates, nuclear charges, and number of electrons) and the orthonormalized basis set, $\{\chi_p\}$.
2. Calculate one- and two-electron integrals: h_{pq}, $\langle pr|qs\rangle$ and $\langle pr|sq\rangle$.
3. Determine the density matrix $\mathbf{P}$ with the initial molecular orbital coefficient, $\{\mathbf{C}_i\}$.
4. Compute the Fock matrix $\mathbf{F}$ in Eq. (2.60).
5. Diagonalize $\mathbf{F}$ to obtain the molecular orbital coefficients, $\{\mathbf{C}_i'\}$, and orbital energies, $\{\epsilon_i\}$.
6. Update the density matrix, $\mathbf{P}'$, with the new molecular orbital coefficients, $\{\mathbf{C}_i'\}$.
7. Compare the given density matrix, $\mathbf{P}'$, with the previous one, $\mathbf{P}$, and halt the SCF process, regarding it as convergent if the difference between these density matrices is less than a given threshold value. If the difference is greater than the threshold value, return to step 4 and continue with the calculation.

This SCF calculation lowers the total electronic energy by the unitary transformation of molecular orbitals. Therefore, this can be taken as a process relaxing molecular orbitals.

Using the density matrix $\mathbf{P}$, the Fock matrix $\mathbf{F}$, and the one-electron integral, the total electronic energy can be calculated by Eqs. (2.54) and (2.57) as

$$E = \frac{1}{2} \sum_{p,q=1}^{n_{\text{basis}}} P_{pq} \left(h_{pq} + F_{pq} \right). \tag{2.68}$$

2.6 Basis Function

In the previous section, the Roothaan method was introduced as a method for speeding up the Hartree–Fock calculation by expanding molecular orbitals with a basis set. The accuracy and computational time of the Roothaan calculations depend on the quality and number of basis functions, respectively. Therefore, it is necessary for reproducing accurate chemical reactions and properties to choose basis functions that give highly accurate molecular orbitals with a minimum number of functions.

The most frequently used basis function is the Gaussian-type function in quantum chemistry calculations. This is due to the following *product rule of Gaussian-type functions* (Szabo and Ostlund 1996).

$$
\begin{aligned}
&\exp(-\alpha|\mathbf{r}-\mathbf{R}_{\mathrm{A}}|^2)\exp(-\beta|\mathbf{r}-\mathbf{R}_{\mathrm{B}}|^2)\\
&\quad=\left[\frac{2\alpha\beta}{(\alpha+\beta)\pi}\right]^{3/4}\exp\left(-\frac{\alpha\beta}{\alpha+\beta}|\mathbf{R}_{\mathrm{A}}-\mathbf{R}_{\mathrm{B}}|^2\right)\exp\left[-(\alpha+\beta)|\mathbf{r}-\mathbf{R}_{\mathrm{C}}|^2\right], \quad (2.69)
\end{aligned}
$$

$$
\mathbf{R}_{\mathrm{C}}=\frac{\alpha}{\alpha+\beta}\mathbf{R}_{\mathrm{A}}+\frac{\beta}{\alpha+\beta}\mathbf{R}_{\mathrm{B}}, \quad (2.70)
$$

where $\mathbf{R}_{\mathrm{A}}$ is the coordinate vector of nucleus A. Based on this rule, the product of Gaussian-type functions is given by a Gaussian-type function centered on a different point, and this drastically speeds up the two-electron integral calculations. However, as shown in the wavefunction of the hydrogen atom (Eq. (1.80)), actual atomic orbitals are close to the Slater-type function, $\exp(-\zeta r)$. It is therefore unfavorable to use Gaussian-type functions as they are. To solve this problem, the *contracted Gaussian-type basis function*, which is a linear combination of Gaussian-type functions, was proposed by Boys (Boys 1950):

$$
\chi_p(\mathbf{r}-\mathbf{R}_{\mathrm{A}})=\sum_{\mu}c_{\mu p}\exp(-\alpha_{\mu p}|\mathbf{r}-\mathbf{R}_{\mathrm{A}}|^2). \quad (2.71)
$$

In this equation, the original Gaussian-type functions are called *primitive functions* to distinguish them from the contracted ones. The primitive functions are specified only by the orbital exponent, $\alpha_{\mu p}$, contracted coefficient, $c_{\mu p}$, and coordinate vector of the center of the function, $\mathbf{R}_{\mathrm{A}}$. By using these contracted Gaussian-type functions as basis functions, it was expected that highly accurate results could be obtained with far fewer basis functions.

Different from the contracted Gaussian-type basis functions mentioned later, primitive functions usually have a standardized form to save the effort of developing a different computational program for each basis function. The primitive functions corresponding to s, p, d, and f atomic orbitals are represented analogically to the spherical harmonic functions (see Sect. 1.8) as

$$
\begin{aligned}
&\text{s function}: \exp(-\alpha r^2),\\
&\text{p function}: (x, y, z)\exp(-\alpha r^2),\\
&\text{d function}: (x^2, y^2, z^2, xy, yz, zx)\exp(-\alpha r^2),\\
&\text{f function}: (x^3, y^3, z^3, x^2y, x^2z, xy^2, y^2z, yz^2, xz^2, xyz)\exp(-\alpha r^2),
\end{aligned}
\tag{2.72}
$$

respectively. Note that the numbers of linear independent d and f functions are essentially five and seven, respectively. Therefore, these primitive functions are generally combined for d and f orbitals to provide linear independent functions: $x^2 - y^2$ function is formed for d orbitals and $x^2z - y^2z$, $3x^2y - y^3$ and $3xy^2 - x^3$ functions are formed for f orbitals.

For contracted Gaussian-type basis functions, various types of functions have been suggested. The following are several major Gaussian-type basis functions (for further details, see, e.g., (Jensen 2006)):

- *Minimal basis functions* (e.g. STO-*L*G) contain only minimal required contracted functions for each atom. For example, since electrons occupy atomic orbitals up to the 2p orbital in the carbon atom, five contracted Gaussian-type functions corresponding to 1s, 2s, $2p_x$, $2p_y$, and $2p_z$ orbitals are necessary at the minimum. The minimal basis functions approximating Slater-type orbitals (STO) corresponding to atomic orbitals with L primitive functions are called *STO-LG basis functions*.
- *Split valence basis functions* (e.g. 6-31G, 6-311G, DZ, TZ, and DZV) use one type of contracted Gaussian-type function for core orbitals and multiple contracted functions for valence orbitals. In most molecules, valence orbitals mainly contribute to chemical bonds, while core orbitals hardly participate in the bonds. It is therefore reasonable to use many basis functions for valence orbitals to calculate electronic states accurately while conserving the total number of basis functions. In this type of basis function, *Pople-type basis functions*, including the 6-31G basis, are included. "6-31" indicates the extent of the contraction and split, in which "6" means the use of contracted basis functions of 6 primitive functions for core orbitals and "31" means the use of doubly-split basis functions combining contracted basis functions of 3 primitive functions with one uncontracted basis function for valence orbitals. "6-311G" uses triply-split basis functions for valence orbitals. In this type of basis function, the *Dunning–Huzinaga-type* and *Ahlrichs-type basis functions* are also included: Dunning–Huzinaga-type functions are described as "DZ," "TZ," and "QZ" for doubly-, triply-, and quadruply-split functions for valence orbitals, while the Ahlrichs-type are similarly written as "DZV," "TZV," and "QZV."
- *Polarization function-supplemented basis functions* (e.g. 6-31G*, 6-31G(d), DZp, cc-pVXZ, and cc-pCVXZ) add *polarization functions* to incorporate the anisotropic nature of molecular orbitals originating from chemical bonds. Polarization functions usually have higher angular momenta than the highest angular momenta of the atomic orbitals that mainly make up the molecular

orbitals. In the Pople-type basis functions, the inclusion of polarization functions is represented by an asterisk "*" such as "6-31G*." For a single asterisk "*," one polarization function is added for all atoms except hydrogen atoms, while for a double asterisk "**," one p orbital function is also added to each hydrogen atom. Recently, the form of polarization function is often written explicitly, for example, "6-31G(d)" (for this basis function, a d orbital function is added). In the case of the Dunning–Huzinaga-type basis functions, polarization functions are represented by "p," for example, "DZp" for a single polarization function and "TZ2p" for double polarization functions. Recent highly accurate calculations frequently use *correlation consistent basis functions*. Dunning et al. developed these basis functions to create the remaining functions so that they produce electron correlations equivalently when they are included in the valence orbitals, such as the 4d and 4f orbitals. These basis functions are represented as "cc-pVXZ" (X=D, T, Q, 5, 6,...). Furthermore, there are "cc-pCVXZ" basis functions, which take into account the electron correlations of the core orbitals by adding s and p orbital functions.

- *Diffuse-function-augmented basis functions* (e.g. 6-311+G(d), 6-311++G(2df, 2pd), and aug-cc-pVXZ) employ *diffuse functions* to take weakly bound electrons into consideration. In particular, these diffuse functions are requisite in the calculations of the anions and excited states of small molecules. Adding diffuse functions is shown by a plus "+" in Pople-type basis functions. "6-311+G(d)" augments sp diffuse functions mixing s and p orbitals for all atoms except hydrogen atoms, and "6-311++G(2df, 2pd)" adds two d and one f diffuse functions for all atoms except hydrogen and two p and one d orbital functions for hydrogen in addition. For correlation consistent basis functions, augmenting diffuse functions is represented by "aug-" at the head of the names and one diffuse function is added to each angular momentum type of basis function. In the "aug-cc-pVDZ" function, containing up to d orbital functions, diffuse functions are added to s through d orbital functions one by one.
- *Effective core potential (ECP) basis functions* (e.g. LanL2DZ and Stuttgart-ECP) approximate core orbitals, which hardly affect reactions and properties in most systems, as effective potentials were introduced to drastically reduce the number of basis functions. In particular, for the fourth-period atoms or later, ECP basis functions are used in most cases, except for the reactions and properties in which core electrons participate, and have given highly accurate results. The most widely used ECP is the *relativistic ECP (RECP)*, incorporating the relativistic effects of the core electrons: The best known ECP basis functions are LanL2DZ, of Los Alamos National Laboratory (USA), and the Stuttgart relativistic small core (STRSC) and large core (STRLC) ECP basis functions (Germany). Among others, there are the *ab initio model potential (AIMP)* basis functions, which were developed to reproduce the ab initio potentials of core orbitals with nodes in order to consider the indirect effects of core electrons.

These are the most frequently used basis functions in quantum chemistry calculations. Of course, there are various basis functions in addition to the above, and some functions have been confirmed to be efficient, especially in calculations for specific systems. Although the accuracies and reliabilities of calculated results clearly depend on the basis functions used, it is usually hard to answer the question which basis set should be used for specific systems. The only way to choose basis functions is to contrast their features.

It is worthwhile to note that although only Gaussian-type basis functions have so far been explained, they are not the only option in quantum chemistry calculations. Actually, Slater-type basis functions, which reproduce the shapes of orbitals more accurately, as mentioned above, are used in, e.g., the Amsterdam Density Functional (ADF) program, in which numerical integral calculations are carried out with Slater-type functions.

Finally, let us examine *basis set superposition error (BSSE)*, which Gaussian-type basis functions generally produce. This error is attributed to the artificial stabilization energy coming from the overlap of nonorthogonal basis functions. Therefore, this error has the effect of making atoms approaching closer to each other. The attracting force from this error is called the *Pulay force* (Pulay 1969). To remove this error, the *counterpoise method* (van Duijneveldt et al. 1994) is frequently used: For example, if system AB is calculated with $a+b$ basis functions, the counterpoise method estimates the BSSE as

$$\Delta E = E_{\mathrm{A}}^{a+b} + E_{\mathrm{B}}^{a+b} + E_{\mathrm{A}}^{a} + E_{\mathrm{B}}^{b}, \tag{2.73}$$

and subtracts this ΔE from the total energy. In this equation, E_{A}^{a+b} indicates the energy of system A calculated with both a and b basis functions. It is well known that BSSE is severe in calculating weak bonds. For example, *van der Waals bonds cannot be reproduced correctly without a BSSE correction* (see Sect. 6.3). However, since the overlap of basis functions is natural in chemical bonds, it is difficult to specify the BSSE in chemical bond calculations. The BSSE is therefore neglected in most calculations involving reactions and properties.

2.7 Coulomb and Exchange Integral Calculations

The Coulomb and exchange integral calculations of the Hartree–Fock equation are usually the rate-determining processes in Hartree–Fock calculations. As seen in Eqs. (2.63) and (2.64), the two-electron integrals called the two-electron repulsion integrals are represented with basis functions and have double integrations in terms of electronic position vectors in three-dimensional space coordinates. These two-electron repulsion integrals are given in the fourth order of basis functions. Since this indicates that the number of integrals drastically increases with the number of basis functions, the integral calculations rapidly approach a limitation on the number of electrons, unless the order is decreased. Fortunately, the two-electron integrals contain many that are non-contributing or duplicate. It is therefore necessary to eliminate these integrals for efficient integral calculations.

In the field of quantum chemistry, many algorithms have been developed to improve the efficiency of the two-electron integral calculations over the years. One of the most frequently used algorithms is the screening based on the *Schwartz* (K.H.A. Schwartz) *inequality* (1888). That is, based on the inequality,

$$\left| \int d^3\mathbf{r} \chi_p^*(\mathbf{r}) \chi_q(\mathbf{r}) \right|^2 \leq \int d^3\mathbf{r} \left| \chi_p(\mathbf{r}) \right|^2 \cdot \int d^3\mathbf{r} \left| \chi_q(\mathbf{r}) \right|^2, \quad (2.74)$$

negligible integrals can be screened out by calculating only the integrals of the squared basis functions. Moreover, various sophisticated algorithms have been suggested for the two-electron repulsion integral calculations. Recent major algorithms combine various existing algorithms: the Obara–Saika recursive algorithm (Obara and Saika 1986) for less-contracted Gaussian-type basis functions, the Rys polynomial algorithm (Dupuis et al. 1976) for large angular momentum basis functions, the Pople–Hehre axis-switching algorithm (Pople and Hehre 1978) for more-contracted basis functions, and so forth. The major example is the PRISM algorithm (Gill and Pople 1991), which is used in Gaussian09 and other programs. By using these algorithms, the order of the two-electron repulsion integral calculations can be decreased to the second to third order for the number of basis functions. However, the calculations of large systems require further decrease of the order.

Coulomb interactions are not quantum mechanical interactions but classical mechanical interactions between charges. Therefore, established techniques for increasing efficiency in classical mechanics are available in Coulomb integral calculations. So far, various techniques for classical mechanical systems have been applied to the Coulomb integral calculations to reduce the computational order. One of the major techniques is the *fast multipole method* (*FMM*) (Greengard and Rokhlin 1987), which has been used to compute the gravity between stars scattered through an astronomical body. FMM distinguishes long-range electron–electron interactions and calculates these interactions by the multipole method and other short-range electron–electron interactions by the usual method. For FMM, there are different methods for distinguishing long-range interactions and expanding multipoles (White et al. 1994). Another major technique for Coulomb integral calculations is the use of the *Poisson equation* (Becke and Dickson 1988; Delley 1996; Manby et al. 2001),

$$\int d^3\mathbf{r}' \frac{\nabla^2 \rho(\mathbf{r}')}{|\mathbf{r} - \mathbf{r}'|} = -4\pi\rho(\mathbf{r}), \quad (2.75)$$

where ρ is the *total electron density* defined by

$$\rho(\mathbf{r}) = 2\sum_i^n |\phi_i(\mathbf{r})|^2. \quad (2.76)$$

In this equation, the Coulomb integral can be transformed into a single integral by replacing the basis functions with the Laplacian of the auxiliary density basis functions $\{\nabla^2 \xi_a\}$. The Coulomb integral is transformed for the combinations of the auxiliary density basis functions as

$$
\begin{aligned}
J_{ab} &= \int d^3\mathbf{r}_1 d^3\mathbf{r}_2 \frac{\left|\nabla_1^2 \xi_a(\mathbf{r}_1)\right| \left|\nabla_2^2 \xi_b(\mathbf{r}_2)\right|}{r_{12}} \\
&= -4\pi \int d^3\mathbf{r} \xi_a(\mathbf{r}) \nabla^2 \xi_b(\mathbf{r}),
\end{aligned} \tag{2.77}
$$

and for the mixed basis functions with the usual basis functions $\{\chi_p\}$ as

$$
\begin{aligned}
J_{ap} &= \int d^3\mathbf{r}_1 d^3\mathbf{r}_2 \frac{\left|\nabla_1^2 \xi_a(\mathbf{r}_1)\right| \left|\chi_p(\mathbf{r}_2)\right|}{r_{12}} \\
&= -4\pi \int d^3\mathbf{r} \xi_a(\mathbf{r}) \chi_p(\mathbf{r}).
\end{aligned} \tag{2.78}
$$

The auxiliary density basis functions $\{\xi_a\}$ are arbitrary electron density basis functions that are smaller than the basis functions $\{\chi_p\}$, spanning the model electron density,

$$
\tilde{\rho}(\mathbf{r}) = \sum_a c_a \xi_a(\mathbf{r}), \tag{2.79}
$$

and determines coefficient $\{c_a\}$ by minimizing

$$
\Delta = \frac{1}{2} \int d^3\mathbf{r}_1 d^3\mathbf{r}_2 \frac{|\rho(\mathbf{r}_1) - \tilde{\rho}(\mathbf{r}_1)| \, |\rho(\mathbf{r}_2) - \tilde{\rho}(\mathbf{r}_2)|}{r_{12}}. \tag{2.80}
$$

This technique is called *electron density fitting* and is employed in the DMol program (Delley 1990). Methods combining the FMM and Poisson methods have also been suggested (Watson et al. 2008). By using these techniques, the order of the Coulomb integral calculations approaches the first order for the number of basis functions. This order can be taken as the order of the electron number N, because the number of basis functions is nearly proportional to the number of electrons. A method that makes a computational process approach the first order of the calculated system is generally called a *linear-scaling method* or *Order-N method*. In the Hartree–Fock method, most linear-scaling methods are proposed for Coulomb integral calculations.

Exchange interactions are purely quantum-mechanical interactions, acting even on distant electrons. Due to these interactions, total electronic energies are lowered, stabilizing electronic states. *A significant proportion of the exchange interactions is taken up by the self-interactions of the electrons themselves.* Exchange self-interactions correspond to the K_{ii} terms in Eq. (2.45) and cancel out with J_{ii} terms, as shown in Eq. (2.43). Consequently, these exchange self-interactions remove the Coulomb potential of one electron. *Exchange interactions increase the overlap of orbitals to produce attractions between orbitals.* As a result, these attractions lead to the delocalization of the electron distribution, which causes the long-range nature of the exchange interactions. Exchange interactions also enhance the quantum nature of electronic states and significantly contribute to the chemical reactions and properties of molecules. Therefore, it is difficult to discuss chemistry even qualitatively without taking exchange interactions into consideration. Similarly to Coulomb integral calculations, various types of linear-scaling methods have been suggested for exchange integral calculations: e.g., the near-field exchange (NFX) method (Burant et al. 1996), based on FMM, and the order-N exchange (ONX) method (Schwegler et al. 1997a), which assumes the exponential decay of the electron density for regions far from given nuclei. However, it has been reported that both methods cannot achieve the linear scaling of exchange integral calculations except for linearly extended long-chain systems. This may be attributed to *the long-range nature of exchange interactions* (Schwegler et al. 1997b) and *the non-classical nature of long-range exchange interactions.*

2.8 Unrestricted Hartree–Fock Method

Thus far, the spin-independent forms of the various calculation methods have been described without clear notification, despite these forms being available only in the electronic state calculations for closed-shell molecules, in which electrons occupy molecular orbitals two by two. To take open-shell molecules containing unpaired electrons into consideration, spin orbitals should be explicitly considered. As a simple method for calculating the electronic states of open-shell molecules, Pople and Nesbet developed the *unrestricted Hartree–Fock (UHF) method* (Pople and Nesbet 1954), which independently deals with the spatial orbitals for α and β spins, in 1954.

In the UHF method, the Hartree–Fock equation in Eq. (2.52) is transformed into simultaneous equations for α- and β-spin electrons,

$$\hat{F}_\alpha \phi_{i\alpha} = \epsilon_{i\alpha}\phi_{i\alpha}, \tag{2.81}$$

and

$$\hat{F}_\beta \phi_{i\beta} = \epsilon_{i\beta}\phi_{i\beta}. \tag{2.82}$$

The spin-dependent Fock operator is given as

$$\hat{F}_\sigma = \hat{h}_\sigma + \sum_j^{n_\sigma}\left(\hat{J}_{j\sigma} - \hat{K}_{j\sigma}\right) + \sum_j^{n_{\sigma'}} \hat{J}_{j\sigma'}, \tag{2.83}$$

where n_σ is the number of σ-spin electrons and $\sigma' \neq \sigma$. Note that each electron has Coulomb interactions but no exchange interactions with opposite-spin electrons. The spin-dependent Coulomb operator $\hat{J}_{j\sigma}$ and exchange operator $\hat{K}_{j\sigma}$ in Eq. (2.83) are defined as

$$\hat{J}_{j\sigma}(\mathbf{r}_1)\phi_{i\sigma}(\mathbf{r}_1) = \int d^3\mathbf{r}_2 \phi^*_{j\sigma}(\mathbf{r}_2)\phi_{j\sigma}(\mathbf{r}_2)\frac{1}{r_{12}}\phi_{i\sigma}(\mathbf{r}_1), \tag{2.84}$$

and

$$\hat{K}_{j\sigma}(\mathbf{r}_1)\phi_{i\sigma}(\mathbf{r}_1) = \int d^3\mathbf{r}_2 \phi^*_{j\sigma}(\mathbf{r}_2)\phi_{i\sigma}(\mathbf{r}_2)\frac{1}{r_{12}}\phi_{j\sigma}(\mathbf{r}_1). \tag{2.85}$$

These equations are called the *UHF equation* or the *Pople–Nesbet equation* (Pople and Nesbet 1954). From the beginning, these equations have the form of the Roothaan method given by simultaneous equations,

$$\mathbf{F}_\alpha \mathbf{C}_{i\alpha} = \epsilon_{i\alpha}\mathbf{S}\mathbf{C}_{i\alpha}, \tag{2.86}$$

and

$$\mathbf{F}_\beta \mathbf{C}_{i\beta} = \epsilon_{i\beta}\mathbf{S}\mathbf{C}_{i\beta}, \tag{2.87}$$

where the spin-dependent Fock operator $\mathbf{F}^\sigma$ and overlap matrix $\mathbf{S}^\sigma$ are given as

$$F_{pq\sigma} = h_{pq\sigma} + \sum_{r,s=1}^{n_{\text{basis}}} P_{sr\sigma}\left(\langle pr|qs\rangle - \langle pr|sq\rangle\right) + \sum_{r,s=1}^{n_{\text{basis}}} P_{pq\sigma'}\langle pr|qs\rangle, \tag{2.88}$$

and

$$S_{pq\sigma} = \int d^3\mathbf{r}\chi^*_{p\sigma}(\mathbf{r})\chi_{q\sigma}(\mathbf{r}). \tag{2.89}$$

Although the solution method of this equation is similar to that of the Roothaan method for closed-shell molecules, it includes pairs of Fock matrices, their diagonalizations, and density matrices for the α- and β-spins.

UHF calculations consequently give different spatial orbitals for α- and β-spins. This causes a problem in that UHF wavefunctions are often not eigenfunctions for the square of the total spin operator, $\hat{\mathbf{S}}^2$. The total spin operator $\hat{\mathbf{S}}$ is the sum of the spin angular momentum operators of the constituent electrons, $\hat{\mathbf{s}}$, and is written as

$$\hat{\mathbf{S}} = (\hat{S}_x, \hat{S}_y, \hat{S}_z) = \sum_i^N \hat{\mathbf{s}}_i. \tag{2.90}$$

Defining the spin functions of molecular orbital ϕ_i as α_i and β_i, the spin angular momentum operator, $\hat{\mathbf{s}} = (\hat{s_x}, \hat{s_y}, \hat{s_z})$, gives

$$\hat{s_z}\begin{pmatrix}\alpha_i\\ \beta_i\end{pmatrix} = \frac{1}{2}\begin{pmatrix}\alpha_i\\ -\beta_i\end{pmatrix},\ \hat{s_x}\begin{pmatrix}\alpha_i\\ \beta_i\end{pmatrix} = \frac{1}{2}\begin{pmatrix}\beta_i\\ \alpha_i\end{pmatrix},\ \hat{s_y}\begin{pmatrix}\alpha_i\\ \beta_i\end{pmatrix} = \frac{i}{2}\begin{pmatrix}\beta_i\\ -\alpha_i\end{pmatrix}, \tag{2.91}$$

and

$$\hat{\mathbf{s}}^2\begin{pmatrix}\alpha_i\\ \beta_i\end{pmatrix} = (\hat{s_x}^2 + \hat{s_y}^2 + \hat{s_z}^2)\begin{pmatrix}\alpha_i\\ \beta_i\end{pmatrix} = \frac{3}{4}\begin{pmatrix}\alpha_i\\ \beta_i\end{pmatrix}. \tag{2.92}$$

Therefore, spin functions are the eigenfunctions of $\hat{s}^2$ and $\hat{s_z}^2$. Since the nonrelativistic Hamiltonian operator generally does not explicitly depend on spins, it is commutative with the $\hat{s}^2$ and $\hat{s_z}$ operators. This indicates that the *total wavefunctions including the spatial functions, Ψ, are the eigenfunctions of $\hat{s}^2$ and $\hat{s_z}^2$*. Therefore, it is easily proven that the Hamiltonian operator is commutative with the square of total spin operator, $\hat{\mathbf{S}}^2$, and the z-component of the total spin operator, $\hat{S}_z$, and consequently, wavefunctions satisfy

$$\hat{\mathbf{S}}^2\Psi = \left(\sum_{i,j}^{N}\hat{\mathbf{s}}_i \cdot \hat{\mathbf{s}}_j\right)\Psi = \left(\frac{N_\alpha - N_\beta}{2}\right)\left(\frac{N_\alpha - N_\beta}{2} + 1\right)\Psi, \tag{2.93}$$

and

$$\hat{S}_z\Psi = \left(\sum_{i}^{N}\hat{s_z}\right)\Psi = \left(\frac{N_\alpha - N_\beta}{2}\right)\Psi, \tag{2.94}$$

where N_σ is the number of σ-spin electrons in molecules. Since the total number of spins is preserved, UHF wavefunctions satisfy Eq. (2.94) to be the eigenfunctions of $\hat{S}_z$. However, since the spatial functions of the UHF wavefunctions are different for spins, the spins for the same number of molecular orbitals do not cancel out, violating Eq. (2.93). Therefore, the *UHF wavefunctions are generally not eigenfunctions of* $\hat{\mathbf{S}}^2$. Actually, for UHF wavefunctions, the expectation value of the $\hat{\mathbf{S}}^2$ operator is proven to be (Szabo and Ostlund 1996)

$$\begin{aligned}\langle\hat{\mathbf{S}}^2\rangle_{\text{UHF}} = &\left(\frac{N_\alpha - N_\beta}{2}\right)\left(\frac{N_\alpha - N_\beta}{2} + 1\right) + N_\beta \\ &- \sum_{i,j}^{N}\left|\int d^3\mathbf{r}\phi_{i\alpha}^*(\mathbf{r})\phi_{j\beta}(\mathbf{r})\right|^2.\end{aligned} \tag{2.95}$$

2.9 Electronic States of Atoms

The electronic configurations in atoms can be explained qualitatively by the Hartree–Fock wavefunctions to some extent. Since atoms are spherically symmetric, potentials are also taken as spherically symmetric. Therefore, similarly to the wavefunction of the hydrogen atom in Eq. (1.78), the angular functions of the atomic wavefunctions are spherical harmonics. On the other hand, the radial functions are different from those for the hydrogen atom, because the degeneracy for the azimuthal quantum number l is resolved due to the appearance of the interelectronic potential, V_{ee}. However, as anticipated from the fact that the eigenenergy of the hydrogen atom in Eq. (1.82) depends only on main quantum number n, the energy differences between the motional states of different l are much smaller than those of different n for relatively light atoms. There are, therefore, possible motional states for natural the numbers l $(l \leq n-1)$ in ascending n number. For the azimuthal quantum number l, the radial functions are represented as s, p, d, f, g, h, ... These radial functions are also degenerate for the magnetic quantum number m, which is an integer $-l \leq m \leq l$. In this way, the electronic motional states, i.e., atomic orbitals, with (n, l, m) quantum numbers are determined as 1s, 2s, 2p ($2p_{-1}$, $2p_0$, $2p_{+1}$), 3s, 3p ($3p_{-1}$, $3p_0$, $3p_{+1}$), 3d ($3d_{-2}$, $3d_{-1}$, $3d_0$, $3d_{+1}$, $3d_{+2}$), ...

How are electrons distributed to these atomic orbitals? The atomic electron configurations are summarized in Tables 2.1–2.3. Electrons occupy the energetically lowest atomic orbitals two by two in closed-shell atoms such as the rare gases. However, it is not so simple in open-shell atoms. As a practical explanation, *Hund's rule* (Hund 1925a,b) is available. This rule consists of the following two rules:

- In electron configurations with the same main and azimuthal quantum numbers, the highest total spin configuration is the most stable.
- In the highest total spin configurations, the highest angular momentum configuration is the most stable.

For example, the electron configuration of the carbon atom is $(1s)^2(2s)^2(2p)^2$, obeying Hund's rule for the assignment of two electrons to the outermost 2p orbitals. That is, 2p orbitals are occupied by two α-spin electrons to maximize the total spin and $2p_{+1}$ and $2p_0$ orbitals are occupied to maximize the total angular momentum. Exchange interactions are the underlying cause for the first rule that the highest total spin configuration is the most stable. The second rule that the highest angular momentum configuration is the most stable is due to Coulomb interactions, which are minimized by allocating electrons close to the equatorial planes of atoms, where electrons are farthest from each other. Actually, atomic ground-state configurations cannot essentially be given by the Hartree method, but can be reproduced by the Hartree–Fock method containing exchange interactions to some extent. Note, however, that electron configurations containing open-shell core orbitals such as those of rare earth atoms cannot be provided without considering electron correlation, as described in the next chapter.

Table 2.1 Electronic configurations of atoms up to krypton

	Atomic		K	L		M			N				O		Ground
Period	number	Element	1s	2s	2p	3s	3p	3d	4s	4p	4d	4f	5s	5p	state
1	1	H	1												$^2S_{1/2}$
	2	He	2												$^1S_0$
2	3	Li	2	1											$^2S_{1/2}$
	4	Be	2	2											$^1S_0$
	5	B	2	2	1										$^2P_{1/2}$
	6	C	2	2	2										$^3P_0$
	7	N	2	2	3										$^4S_{3/2}$
	8	O	2	2	4										$^3P_2$
	9	F	2	2	5										$^2P_{3/2}$
	10	Ne	2	2	6										$^1S_0$
3	11	Na	2	2	6	1									$^2S_{1/2}$
	12	Mg	2	2	6	2									$^1S_0$
	13	Al	2	2	6	2	1								$^2P_{1/2}$
	14	Si	2	2	6	2	2								$^3P_0$
	15	P	2	2	6	2	3								$^4S_{3/2}$
	16	S	2	2	6	2	4								$^3P_2$
	17	Cl	2	2	6	2	5								$^2P_{3/2}$
	18	Ar	2	2	6	2	6								$^1S_0$
4	19	K	2	2	6	2	6		1						$^2S_{1/2}$
	20	Ca	2	2	6	2	6		2						$^1S_0$
	21	Sc	2	2	6	2	6	1	2						$^2D_{3/2}$
	22	Ti	2	2	6	2	6	2	2						$^3F_2$
	23	V	2	2	6	2	6	3	2						$^4F_{3/2}$
	24	Cr	2	2	6	2	6	5	1						$^7S_3$
	25	Mn	2	2	6	2	6	5	2						$^6S_{5/2}$
	26	Fe	2	2	6	2	6	6	2						$^5D_4$
	27	Co	2	2	6	2	6	7	2						$^4F_{9/2}$
	28	Ni	2	2	6	2	6	8	2						$^3F_4$
	29	Cu	2	2	6	2	6	10	1						$^2S_{1/2}$
	30	Zn	2	2	6	2	6	10	2						$^1S_0$
	31	Ga	2	2	6	2	6	10	2	1					$^2P_{1/2}$
	32	Ge	2	2	6	2	6	10	2	2					$^3P_0$
	33	As	2	2	6	2	6	10	2	3					$^4S_{3/2}$
	34	Se	2	2	6	2	6	10	2	4					$^3P_2$
	35	Br	2	2	6	2	6	10	2	5					$^2P_{3/2}$
	36	Kr	2	2	6	2	6	10	2	6					$^1S_0$

In addition, there is a third, complementary rule:

- The configuration with the lowest total angular momentum (total spin + total orbital angular momentum) is the most stable for atoms containing less-than-half occupied outermost orbitals, and the configuration with the highest total angular momentum is the most stable for those containing greater-than-half occupied outermost orbitals.

Table 2.2 Electronic configurations of atoms from rubidium to lutetium

	Atomic		K 1s	N		O				P				Ground
Period	number	Element	- N 4p	4d	4f	5s	5p	5d	5f	6s	6p	6d	6f	state
5	37	Rb	[Kr]			1								$^2S_{1/2}$
	38	Sr	[Kr]			2								$^1S_0$
	39	Y	[Kr]	1		2								$^2D_{3/2}$
	40	Zr	[Kr]	2		2								$^3F_2$
	41	Nb	[Kr]	4		1								$^6D_{1/2}$
	42	Mo	[Kr]	5		1								$^7S_3$
	43	Tc	[Kr]	6		1								$^6S_{5/2}$
	44	Ru	[Kr]	7		1								$^5F_5$
	45	Rh	[Kr]	8		1								$^4F_{9/2}$
	46	Pd	[Kr]	10										$^1S_0$
	47	Ag	[Kr]	10		1								$^2P_{1/2}$
	48	Cd	[Kr]	10		2								$^3P_0$
	49	In	[Kr]	10		2	1							$^4S_{3/2}$
	50	Sn	[Kr]	10		2	2							$^3P_2$
	51	Sb	[Kr]	10		2	3							$^2P_{3/2}$
	52	Te	[Kr]	10		2	4							$^1S_0$
	53	I	[Kr]	10		2	5							$^2S_{1/2}$
	54	Xe	[Kr]	10		2	6							$^1S_0$
6	55	Cs	[Kr]	10		2	6			1				$^2D_{3/2}$
	56	Ba	[Kr]	10		2	6			2				$^3H_4$
	57	La	[Kr]	10		2	6	1		2				^4I
	58	Ce	[Kr]	10	2	2	6			2				$^5I_4$
	59	Pr	[Kr]	10	3	2	6			2				^6H
	60	Nd	[Kr]	10	4	2	6			2				$^7F_0$
	61	Pm	[Kr]	10	5	2	6			2				$^8S_{7/2}$
	62	Sm	[Kr]	10	6	2	6			2				$^9D_2$
	63	Eu	[Kr]	10	7	2	6			2				$^6H_{15/2}$
	64	Gd	[Kr]	10	7	2	6	1		2				^5I
	65	Tb	[Kr]	10	9	2	6			2				^4I
	66	Dy	[Kr]	10	10	2	6			2				$^3H_4$
	67	Ho	[Kr]	10	11	2	6			2				$^2F_{7/2}$
	68	Er	[Kr]	10	12	2	6			2				$^1S_0$
	69	Tm	[Kr]	10	13	2	6			2				$^2D_{3/2}$
	70	Yb	[Kr]	10	14	2	6			2				$^3F_2$
	71	Lu	[Kr]	10	14	2	6	1		2				$^4F_{3/2}$

"[Kr]" indicates the electron configuration of krypton

This rule is based on *spin–orbit interactions*, which are based on relativistic effects (see Sect. 6.4). For example, let us compare the electron configurations of silicon and sulfur atoms. Since the outermost 3p orbitals are occupied by two and four electrons in silicon and sulfur atoms, respectively, the silicon atom, containing less-than-half occupancy, has a $^3P_0$ ground state with the minimum zero total angular

Table 2.3 Electronic configurations of atoms from hafnium to rutherfordium

	Atomic		K 1s	N		O				P				Q		Ground
Period	number	Element	- N 4p	4d	4f	5s	5p	5d	5f	6s	6p	6d	6f	7s	7p	state
6	72	Hf	[Kr]	10	14	2	6	2		2						$^5D_0$
	73	Ta	[Kr]	10	14	2	6	3		2						$^6S_{5/2}$
	74	W	[Kr]	10	14	2	6	4		2						$^5D_4$
	75	Re	[Kr]	10	14	2	6	5		2						$^4F_{9/2}$
	76	Os	[Kr]	10	14	2	6	6		2						$^3D_3$
	77	Ir	[Kr]	10	14	2	6	7		2						$^2S_{1/2}$
	78	Pt	[Kr]	10	14	2	6	9		1						$^1S_0$
	79	Au	[Kr]	10	14	2	6	10		1						$^2P_{1/2}$
	80	Hg	[Kr]	10	14	2	6	10		2						$^3P_0$
	81	Tl	[Kr]	10	14	2	6	10		2	1					$^4S_{3/2}$
	82	Pb	[Kr]	10	14	2	6	10		2	2					$^3P_2$
	83	Bi	[Kr]	10	14	2	6	10		2	3					$^2P_{3/2}$
	84	Po	[Kr]	10	14	2	6	10		2	4					$^1S_0$
	85	At	[Kr]	10	14	2	6	10		2	5					$^2S_{1/2}$
	86	Rn	[Kr]	10	14	2	6	10		2	6					$^1S_0$
7	87	Fr	[Kr]	10	14	2	6	10		2	6			1		$^2D_{3/2}$
	88	Ra	[Kr]	10	14	2	6	10		2	6			2		$^3F_2$
	89	Ac	[Kr]	10	14	2	6	10		2	6	1		2		$^2D_{3/2}$
	90	Th	[Kr]	10	14	2	6	10		2	6	2		2		$^3F^a$
	91	Pa	[Kr]	10	14	2	6	10	2	2	6	1		2		$^4K^a$
	92	U	[Kr]	10	14	2	6	10	3	2	6	1		2		$^5L^a$
	93	Np	[Kr]	10	14	2	6	10	4	2	6	1		2		$^6L^a$
	94	Pu	[Kr]	10	14	2	6	10	5	2	6	1		2		$^7K^a$
	95	Am	[Kr]	10	14	2	6	10	7	2	6			2		$^8S^a$
	96	Cm	[Kr]	10	14	2	6	10	7	2	6	1		2		$^9D^a$
	97	Bk	[Kr]	10	14	2	6	10	8	2	6	1		2		$^8H^a$
	98	Cf	[Kr]	10	14	2	6	10	10	2	6			2		$^5I^a$
	99	Es	[Kr]	10	14	2	6	10	11	2	6			2		$^4I^a$
	100	Fm	[Kr]	10	14	2	6	10	12	2	6			2		$^3H^a$
	101	Md	[Kr]	10	14	2	6	10	12	2	6	1		2		$^4K^a$
	102	No	[Kr]	10	14	2	6	10	14	2	6			2		$^1S^a$
	103	Lr	[Kr]	10	14	2	6	10	14	2	6			2	1	$^2D^a$
	104	Rf	[Kr]	10	14	2	6	10	14	2	6	2		2		$^3F_2{}^a$

Ground states designated by "*a*" (Küchle et al. 1994) and "*b*" (Eliav et al. 1995) are the most stable electronic states, and their electron configurations in relativistic full-electron calculations. "[Kr]" indicates the electron configuration of krypton

momentum and the sulfur atom, containing greater-than-half occupancy, has a $^3P_2$ ground state with the maximum two total angular momentum. It is required to include the relativistic spin–orbit interactions to reproduce these ground states.

References

Becke, A.D., Dickson, R.M.: J. Chem. Phys. **89**, 2993–2997 (1988)
Born, M., Oppenheimer, R.: Ann. Phys. **389**, 457–484 (1927)
Boys, S.F.: Proc. R. Soc. Lond. A. **200**, 542–554 (1950)
Burant, J.C., Scuseria, G.E., Frisch, M.J.: J. Chem. Phys. **105**, 8969–8972 (1996)
Coulson, C.A.: Proc. Camb. Philos. Soc. **34**, 204–212 (1938)
Delley, B.: J. Chem. Phys. **92**, 508–517 (1990)
Delley, B.: J. Phys. Chem. **100**, 6107–6110 (1996)
Dirac, P.A.M.: Proc. R. Soc. Lond. A **112**, 661–677 (1926)
Dupuis, M., Rys, J., King, H.F.: J. Chem. Phys. **65**, 111–116 (1976)
Ekeland, I.: Le Meilleur des Mondes Possibles (Japanese). Editions du Seuil (2009)
Eliav, E., Kaldor, U., Ishikawa, Y.: Phys. Rev. Lett. **74**, 1079–1082 (1995)
Fock, V.: Z. Phys. **61**, 126–148 (1930)
Fukui, K., Yonezawa, T., Shingu, H.: J. Chem. Phys. **20**, 722–725 (1952)
Gill, P.M.W., Pople, J.A.: Int. J. Quantum Chem. **40**, 753–772 (1991)
Greengard, L., Rokhlin, V.: J. Comput. Phys. **73**, 325–348 (1987)
Hall, G.G.: Proc. R. Soc. Lond. A **205**, 541–552 (1951)
Hartree, D.R.: Math. Proc. Camb. Philos. Soc. **24**, 89–132; 426–437 (1928)
Heisenberg, W.: Z. Phys. **38**, 411–426 (1926)
Hund, F.: Z. Phys. **33**, 345–374 (1925a)
Hund, F.: Z. Phys. **34**, 296–308 (1925b)
Hund, F.: Z. Phys. **36**, 657–674 (1926)
Jensen, F.: Introduction to Computational Chemistry. Wiley, Chichester (2006)
Küchle, W., Dolg, M., Stoll, H., Preuss, H.: J. Chem. Phys. **100**, 7535–7542 (1994)
Landau, L.D.: Z. Phys. **45**, 430–441 (1927)
Lennard-Jones, J.E.: Trans. Faraday Soc. **25**, 668–676 (1929)
Manby, F.R., Knowles, P.J., Lloyd, A.W.: J. Chem. Phys. **115**, 9144–9148 (2001)
Mulliken, R.S.: Phys. Rev. **29**, 637–649 (1927)
Obara, S., Saika, A.: J. Chem. Phys. **84**, 3963–3974 (1986)
Pauli, W.: Z. Phys. **31**, 765–783 (1925)
Pople, J.A., Hehre, W.J.: J. Comput. Chem. **27**, 161–168 (1978)
Pople, J.A., Nesbet, R.K.: J. Chem. Phys. **22**, 571–572 (1954)
Pulay, P.: Mol. Phys. **17**, 197–204 (1969)
Roothaan, C.C.J.: Rev. Mod. Phys. **23**, 69–89 (1951)
Schwegler, E., Challacombe, M., Head-Gordon, M.: J. Chem. Phys. **106**, 9703–9707 (1997a)
Schwegler, E., Challacombe, M., Head-Gordon, M.: J. Chem. Phys. **106**, 9708–9717 (1997b)
Slater, J.C.: Phys. Rev. **34**, 1293–1322 (1929)
Slater, J.C.: Phys. Rev. **35**, 210–211 (1930)
Szabo, A., Ostlund, N.S.: Modern Quantum Chemistry Introduction to Advanced Electronic Structure Theory. Dover, New York (1996)
van Duijneveldt, F.B., van Duijneveldt-van de Rijdt, J.G.C.M., van Lenthe, J.H.: Chem. Rev. **94**, 1873–1885 (1994)
von Neumann, J.: Göttinger Nachr. **1**, 245–272 (1927)
Watson, M.A., Kurashige, Y., Nakajima, T., Hirao, K.: J. Chem. Phys. **128**, 054105(1–7) (2008)
White, C.A., Johnson, B.G., Gill, P.M.W., Head-Gordon, M.: Chem. Phys. Lett. **230**, 8–16 (1994)

Chapter 3
Electron Correlation

3.1 Electron Correlation

Although the Hartree–Fock method is a milestone of modern quantum chemistry, it has gradually become obvious that this method has problems with chemical calculations of various molecules. For example, the Hartree–Fock method is not suitable for quantitative discussions even of simple chemical reactions. Quantitative discussions of actual chemistry need *chemical accuracy*, with errors of about 0.1 Å for bond distances and several kcal/mol for energies, which cannot be attained by Hartree–Fock calculations. This problem is attributed to the energy differences neglected in Hartree–Fock calculations, which are only on the order of 1% of the total energies. In 1955, Löwdin defined this energy difference as the *electron correlation* (Löwdin 1955). That is, electron correlation is defined as the difference between the exact and Hartree–Fock energies,

$$E_{\mathrm{corr}} = E_{\mathrm{exact}} - E_{\mathrm{HF}}. \tag{3.1}$$

So, why is electron correlation produced? This is clearly explained by considering the density matrix. Instead of the density matrix of electrons in molecular orbital spaces given in Eq. (2.65), let us consider the density matrix of electrons in a three-dimensional space. In this case, the first-order density matrix is defined for the wavefunctions of N electron systems $\Psi(\mathbf{r}_1, \mathbf{r}_2, \ldots, \mathbf{r}_N)$ as

$$P(\mathbf{r}'_1, \mathbf{r}_1) = N \int d^3\mathbf{r}_2 \cdots d^3\mathbf{r}_N \Psi(\mathbf{r}'_1, \mathbf{r}_2, \ldots, \mathbf{r}_N)\Psi^*(\mathbf{r}_1, \mathbf{r}_2, \ldots, \mathbf{r}_N), \tag{3.2}$$

where $P(\mathbf{r}_1, \mathbf{r}_1)d^3\mathbf{r}_1$ is the probability of finding out one electron in a volume element $d^3\mathbf{r}_1$, which is integrated over the whole space to the number of electrons. Similarly, the second-order density matrix is defined as

T. Tsuneda, *Density Functional Theory in Quantum Chemistry*,
DOI 10.1007/978-4-431-54825-6_3,

$$\Pi(\mathbf{r}_1', \mathbf{r}_2'; \mathbf{r}_1, \mathbf{r}_2) = \frac{N(N-1)}{2} \int d^3\mathbf{r}_3 \cdots d^3\mathbf{r}_N \Psi(\mathbf{r}_1', \mathbf{r}_2', \ldots, \mathbf{r}_N) \Psi^*(\mathbf{r}_1, \mathbf{r}_2, \ldots, \mathbf{r}_N), \quad (3.3)$$

where $\Pi(\mathbf{r}_1, \mathbf{r}_2; \mathbf{r}_1, \mathbf{r}_2) d^3\mathbf{r}_1 d^3\mathbf{r}_2$ corresponds to the probability of finding one electron in $d^3\mathbf{r}_1$ and another electron in $d^3\mathbf{r}_2$ simultaneously. The time-independent Schrödinger equation for electronic motions in molecules is written as

$$\hat{H}\Psi = \left(-\sum_i^N \frac{\nabla_i^2}{2} + V_{\text{ne}} + \sum_{i<j}^N \frac{1}{r_{ij}} \right) \Psi = E\Psi. \quad (3.4)$$

In this equation, the Hamiltonian contains the $1/r_{ij}$ term, giving infinity for $r_{ij} = 0$, despite the fact that the energy E has a finite value. Therefore, the wavefunction Ψ annihilates at $r_{ij} = 0$,

$$\Pi(\mathbf{r}_1, \mathbf{r}_2; \mathbf{r}_1, \mathbf{r}_2)_{r_{12}\to 0} \Rightarrow \Pi(\mathbf{r}_1, \mathbf{r}_1; \mathbf{r}_1, \mathbf{r}_1) = 0. \quad (3.5)$$

Note that based on the Pauli exclusion principle, this is not applicable to electron pairs with the same quantum numbers except that of spin. This indicates that the wavefunction Ψ is considered to have a hole at $r_{12} = 0$, which is called a *Coulomb hole*. By excluding close electrons, this Coulomb hole decreases the Coulomb interactions, thus lowering the total energies. This energy gain corresponds to the electron correlation.

Next, let us focus on the Hartree–Fock wavefunction at $r_{ij} = 0$ by considering the Slater determinant for the electronic state of the He atom given in Eq. (2.32),

$$\Phi_{\text{HF}}(\mathbf{r}_1, \mathbf{r}_2) = \frac{1}{\sqrt{2}} \left[\phi_1(\mathbf{r}_1)\phi_2(\mathbf{r}_2) - \phi_1(\mathbf{r}_2)\phi_2(\mathbf{r}_1) \right]. \quad (3.6)$$

Note, however, that spin functions are not explicitly taken into consideration. In the case that atomic orbitals, ϕ_1 and ϕ_2, have parallel spins ($\sigma_1 = \sigma_2$), the Hartree–Fock wavefunction approaches

$$\Phi_{\text{HF}}(\mathbf{r}_1, \mathbf{r}_2)_{r_{12}\to 0} \Rightarrow \frac{1}{\sqrt{2}} \left[\phi_1(\mathbf{r}_1)\phi_2(\mathbf{r}_1) - \phi_1(\mathbf{r}_1)\phi_2(\mathbf{r}_1) \right] = 0. \quad (3.7)$$

Therefore, by Eq. (3.3), the diagonal terms of the parallel-spin second-order density matrix are given in the limit of $r_{12} \to 0$ as

$$\Pi_{\text{HF}}^{\sigma_1=\sigma_2}(\mathbf{r}_1, \mathbf{r}_2; \mathbf{r}_1, \mathbf{r}_2)_{r_{12}\to 0} \Rightarrow \Pi_{\text{HF}}^{\sigma_1=\sigma_2}(\mathbf{r}_1, \mathbf{r}_1; \mathbf{r}_1, \mathbf{r}_1) = 0. \quad (3.8)$$

This indicates that the Hartree–Fock wavefunction has a hole for parallel-spin electron pairs. This hole is called the *Fermi hole* or *exchange hole*. In contrast, for anti-parallel spins ($\sigma_1 \neq \sigma_2$), the Hartree–Fock wavefunction approaches

$$\Pi_{\rm HF}^{\sigma_1 \neq \sigma_2}(\mathbf{r}'_1, \mathbf{r}'_2; \mathbf{r}_1, \mathbf{r}_2) = \Phi_{\rm HF}(\mathbf{r}'_1, \mathbf{r}'_2)\Phi^*_{\rm HF}(\mathbf{r}_1, \mathbf{r}_2)$$
$$= \frac{1}{2}\left[P(\mathbf{r}'_1, \mathbf{r}_1)P(\mathbf{r}'_2, \mathbf{r}_2) - P(\mathbf{r}'_1, \mathbf{r}_2)P(\mathbf{r}'_2, \mathbf{r}_1)\right]. \quad (3.9)$$

For $\sigma_1 \neq \sigma_2$, the second term of the right-hand side is zero by Eq. (2.41). The diagonal terms of the anti-parallel-spin second-order density matrix at $r_{12} \to 0$ are, therefore, provided as

$$\Pi_{\rm HF}^{\sigma_1 \neq \sigma_2}(\mathbf{r}_1, \mathbf{r}_2; \mathbf{r}_1, \mathbf{r}_2)_{r_{12}\to 0} = \frac{1}{2}P(\mathbf{r}_1, \mathbf{r}_1)P(\mathbf{r}_2, \mathbf{r}_2)\Big|_{\mathbf{r}_2 \to \mathbf{r}_1}. \quad (3.10)$$

This indicates that, for electron pairs of different spins, electronic motions are independent and have no correlation with each other. Obviously, it is not consistent with the above discussion that wavefunctions have Coulomb holes for close electrons. The Coulomb holes for anti-parallel-spin electron pairs are the main cause for electron correlation excluded in the Hartree–Fock wavefunction.

3.2 Dynamical and Nondynamical Correlations

The Coulomb hole that wavefunctions are predicted to have for close anti-parallel-spin electrons is also called a *correlation hole*. As a condition for wavefunctions containing correlation holes, Kato proposed a *correlation cusp condition* (Kato 1957),

$$\left(\frac{\partial \Psi}{\partial r_{12}}\right)_{r_{12}=0} = \left(\frac{\Psi}{2}\right)_{r_{12}=0}. \quad (3.11)$$

The Hartree–Fock wavefunction violates this condition, because it gives zero for the left-hand side of this equation. As shown in Fig. 3.1, a wavefunction satisfying this condition contains a correlation hole, which contains a sharp dip, called a *cusp*, near $r_{12} = 0$. This correlation hole causes anti-parallel-spin electrons to be further apart, and therefore reduces Coulomb interactions, thus lowering the total electronic energies. Sinanoğlu named this electron correlation in the correlation cusp condition as *dynamical correlation* (Sinanoğlu 1964).

How does dynamical correlation stabilize electronic states? This is explained by considering the energy of the *Møller–Plesset perturbation method*, which is a perturbation method based on the Hartree–Fock wavefunction (McWeeny 1992). In this method, by assuming the sum of the Fock operators as the nonperturbative operator, the Hamiltonian is defined as

$$\hat{H} = \hat{H}_0 + V = \sum_i^n \hat{F}_i + V. \quad (3.12)$$

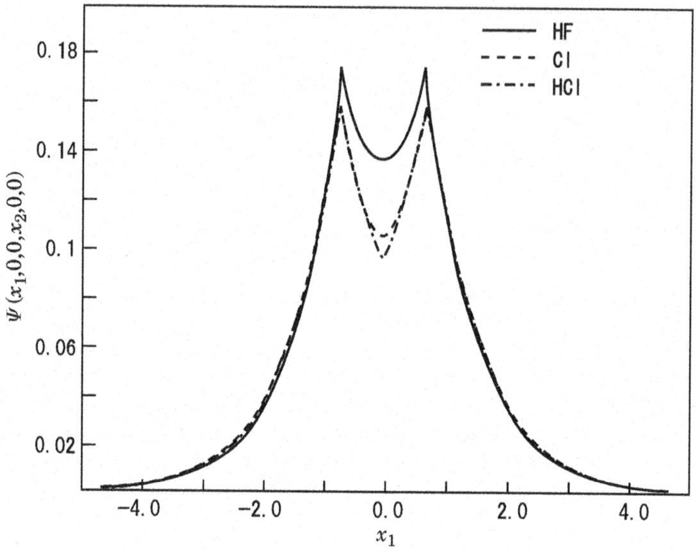

Fig. 3.1 Correlation cusp in the wavefunction of the hydrogen molecule (Frye et al. 1990). The wavefunctions of the Hartree–Fock and configuration interaction (CI) (see Sect. 3.3) methods are compared to the wavefunction of the Hylleraas CI (HCI) method (see Sect. 3.5), which is close to the exact wavefunction. Since x_2 is set to zero in this figure, x_1 corresponds to r_{12}. Note that the wavefunction is not zero even for $r_{12} = 0$, because there remains one electron at $r_{12} = 0$

Consequently, the second-order perturbation electron correlation for the Hartree–Fock energy is given as (Møller and Plesset 1934)

$$E_{\mathrm{MP2}} = -\sum_{i<j}^{n_{\mathrm{occ}}} \sum_{a<b}^{n_{\mathrm{vir}}} \frac{|\langle ij|ab\rangle - \langle ij|ba\rangle|^2}{\epsilon_a + \epsilon_b - \epsilon_i - \epsilon_j}. \tag{3.13}$$

Note that the first-order perturbation electron correlation is always zero in this method (see Sect. 3.4). This second-order perturbation energy originates from the relaxation of electron pairs in orbitals. That is, it is a dynamical correlation and is called a *electron-pair correlation*. What is important is that the Coulomb overlaps (i.e., the overlaps through $1/r_{12}$) of occupied orbitals, ϕ_i and ϕ_j, and of virtual orbitals, ϕ_a and ϕ_b, inversely relate to the differences between integrals $\langle ij|ab\rangle$ and $\langle ij|ba\rangle$, and consequently the numerator of Eq. (3.13). That is, electron-pair correlations are expected to be large for, e.g., $(\mathrm{n}, \pi) \rightarrow (\mathrm{n}^*, \pi^*)$ and $(1\mathrm{s}, 2\mathrm{p}) \rightarrow (1\mathrm{s}^*, 2\mathrm{p}^*)$. Consequently, electron-pair correlation decreases the Coulomb overlap of orbitals ϕ_i and ϕ_j, while exchange interaction contrastingly increases the Coulomb overlap of these orbitals. It is therefore supposed that *dynamical electron correlation provides an energetical stabilization by decreasing Coulomb interactions between occupied orbitals and between virtual orbitals through electron-pair relaxations*. For example, for the helium atom, which contains two electrons, thus giving only this electron-pair correlation, the electron correlation energy is −0.0420 atomic units, which is about 4% of the exchange energy, −1.026 atomic units. This percentage is higher than those of other systems. In most systems, the *electron correlation energy is around 2–3% of the exchange energy.*

Besides dynamical electron correlation, there is *nondynamical electron correlation*, which originates from the explicit interactions between near-degenerate electron configurations (Sinanoğlu 1964). *Nondynamical electron correlation appears most prominently in the dissociations of chemical bonds.* Let us consider the bond dissociation of the hydrogen molecule in its ground state. In the ground state given by the Hartree–Fock method, two electrons occupy the molecular orbital,

$$\phi_1 = \frac{1}{\sqrt{2+2S_{\mathrm{AB}}}}\left(\chi_{\mathrm{A}} + \chi_{\mathrm{B}}\right), \tag{3.14}$$

where χ_{A} and χ_{B} are the atomic orbitals of the hydrogen atoms A and B, and S_{AB} is the overlap integral of the atomic orbitals,

$$S_{\mathrm{AB}} = \int d^3\mathbf{r}\chi_{\mathrm{A}}^*(\mathbf{r})\chi_{\mathrm{B}}(\mathbf{r}). \tag{3.15}$$

In this case, the wavefunction of the hydrogen molecule is represented as the Slater determinant as

$$\Phi_{\mathrm{grd}}(\mathbf{r}_1, \mathbf{r}_2) = \frac{1}{\sqrt{2}} \det\left|\phi_1(\mathbf{r}_1)\phi_1(\mathbf{r}_2)\right|. \tag{3.16}$$

However, at the dissociation limit of $H_2 \rightarrow 2H$, this wavefunction becomes

$$\begin{aligned}\Phi_{\mathrm{grd}}(\mathbf{r}_1, \mathbf{r}_2) = \frac{1}{2\sqrt{2}} \det|&\chi_{\mathrm{A}}(\mathbf{r}_1)\chi_{\mathrm{B}}(\mathbf{r}_2) + \chi_{\mathrm{B}}(\mathbf{r}_1)\chi_{\mathrm{A}}(\mathbf{r}_2)\\ &+\chi_{\mathrm{A}}(\mathbf{r}_1)\chi_{\mathrm{A}}(\mathbf{r}_2) + \chi_{\mathrm{B}}(\mathbf{r}_1)\chi_{\mathrm{B}}(\mathbf{r}_2)|.\end{aligned} \tag{3.17}$$

In this equation, the overlap integral S_{AB} is assumed to be zero at the dissociation limit. Note that the latter two terms in this wavefunction are ionically bonded states, in which electrons are biased to one atom, and therefore, these terms are much less stable than the former two terms corresponding to covalently bonded states at the dissociation limit. Actually, in the restricted Hartree–Fock method, the ground state of the hydrogen molecule is quite unstable at the dissociation limit. This problem is solved by linearly coupling the excited electron configuration of the anti-bonding molecular orbital. Let us consider the Slater determinant of an excited electron configuration,

$$\Phi_{\mathrm{exc}}(\mathbf{r}_1, \mathbf{r}_2) = \frac{1}{\sqrt{2}} \det\left|\phi_2(\mathbf{r}_1)\phi_2(\mathbf{r}_2)\right|, \tag{3.18}$$

which is represented using the anti-bonding molecular orbital,

$$\phi_2 = \frac{1}{\sqrt{2 - 2S_{AB}}} (\chi_A - \chi_B) . \tag{3.19}$$

At the dissociation limit, this electron configuration gives

$$\begin{aligned}\Phi(\mathbf{r}_1, \mathbf{r}_2) = & -\frac{1}{2\sqrt{2}} \det |\chi_A(\mathbf{r}_1)\chi_B(\mathbf{r}_2) + \chi_B(\mathbf{r}_1)\chi_A(\mathbf{r}_2) \\ & -\chi_A(\mathbf{r}_1)\chi_A(\mathbf{r}_2) - \chi_B(\mathbf{r}_1)\chi_B(\mathbf{r}_2)| .\end{aligned} \tag{3.20}$$

Note that the signs of the ionically bonded states are reverse to those of the covalently bonded states.Therefore, the linear combination of the Slater determinantsof the ground and excited electron configurations cancels out the unstable ionically bonded states in the wavefunction at the dissociation limit,

$$\begin{aligned}\Psi(\mathbf{r}_1, \mathbf{r}_2) &= \frac{1}{2\sqrt{2}} \det |\phi_1(\mathbf{r}_1)\phi_1(\mathbf{r}_2) - \phi_2(\mathbf{r}_1)\phi_2(\mathbf{r}_2)| \\ &= \frac{1}{\sqrt{2}} \det |\chi_A(\mathbf{r}_1)\chi_B(\mathbf{r}_2) + \chi_B(\mathbf{r}_1)\chi_A(\mathbf{r}_2)| .\end{aligned} \tag{3.21}$$

In general, electronic states are represented as the linear combination of electron configurations, Φ_I, which are given by the Slater determinants, such as

$$\Psi(\mathbf{r}_1, \mathbf{r}_2, \ldots) = \sum_I C_I \Phi_I(\mathbf{r}_1, \mathbf{r}_2, \ldots). \tag{3.22}$$

These C_I coefficients, which are called CI coefficients, can accept a value of neither 0 nor 1. Actually, electron configurations of both the same spatial and spin symmetries are mixed, if these configurations have close energies. The stabilization energy gained from the mixing of these near-degenerate electron configurations is called *nondynamical electron correlation*, while the functions of these electron configurations are called *configuration state functions (CSFs)*.

3.3 Configuration Interaction

As shown in the previous section, electron correlation can be essentially incorporated by a method that linearly combines excited CSFs with the ground CSF. This method is called the *configuration interaction (CI) method* (McWeeny 1992). Slater indicated the lack of CI in the Hartree–Fock method, before the Hartree–Fock SCF method was developed. In his paper published in 1929, in which he described how the formulation of the Hartree–Fock method is derived with the Slater determinant, he pointed out the problem of exchanging only occupied orbital electrons in the Slater determinants of wavefunctions and suggested a CI method that would take the exchanges with unoccupied orbital electrons into account (Slater 1929). Condon developed the CI method by proposing the *Slater–Condon rule* for the set of contributing electron configurations (Condon 1930). In the CI method, the wavefunction is represented as the linear combination of the Slater determinants for the Hartree–Fock wavefunction as the ground CSF and the excited CSFs of single excitations $\Phi_{i\to a}$, double excitations $\Phi_{ij\to ab}$, and so on,

$$\Psi(\{\mathbf{x}\}) = C_{\mathrm{HF}}\Phi_{\mathrm{HF}}(\{\mathbf{x}\}) + \sum_{i}^{n_{\mathrm{occ}}}\sum_{a}^{n_{\mathrm{vir}}} C_{i\to a}\Phi_{i\to a}(\{\mathbf{x}\}) + \sum_{i,j}^{n_{\mathrm{occ}}}\sum_{a,b}^{n_{\mathrm{vir}}} C_{ij\to ab}\Phi_{ij\to ab}(\{\mathbf{x}\}) + \cdots, \quad (3.23)$$

where $\{\mathbf{x}\}$ is a spin-space coordinate vector $\{\{\mathbf{r}\},\{\sigma\}\}$. Based on this CI wavefunction, the CI Hamiltonian matrix elements,

$$H_{IJ} = \int d^3\{\mathbf{x}\}\Phi_I^*(\{\mathbf{x}\})\hat{H}\Phi_J(\{\mathbf{x}\}), \quad (3.24)$$

are calculated, and then the CI determinant,

$$|\mathbf{H} - E\mathbf{I}| = 0, \quad (3.25)$$

where $\mathbf{I}$ is the unit matrix, is solved by diagonalizing the Hamiltonian matrix $\mathbf{H}$. Since the *full-CI method* incorporating all electron configurations can include all possible electron correlations, exact calculations can be performed by this method, in theory. However, the exactness is practically limited by the finite number of basis functions. Full-CI calculations are also unrealistic except for quite small molecules, because the number of CSFs increases exponentially with respect to the number of electrons. To solve this problem, CI methods that take into account only the excitations of a few electrons, e.g., CI singles and doubles (SDCI) method, have been developed. However, these methods violate *size-consistency* (Pople et al. 1977), which establishes that the energy of a system should equal the energy sum of the separate subsystems at the bond dissociation limit.

The *multiconfigurational SCF (MCSCF) method* (Frenkel 1934) is one of the CI methods that have been developed to efficiently reduce the number of CSFs. In the MCSCF method, the CI calculations are carried out using CSFs restricted to the exchanges of valence orbitals, and then the SCF calculations are performed using the given CI wavefunctions to provide the electronic structures (McWeeny 1992). The MCSCF wavefunction is represented as the linear combination of the CSFs generated from the *active space*, which is spanned by the electron configurations of excitations from specific occupied to virtual orbitals and the ground configuration,

$$\Psi = \sum_{I}^{n_{\mathrm{active\ CSF}}} C_I\Phi_I. \quad (3.26)$$

Using this wavefunction, the MCSCF equation,

$$\sum_s \hat{F}_{rs}\phi_s = \sum_s \epsilon_{sr}\phi_s \tag{3.27}$$

is solved instead of the Hartree–Fock equation in Eq. (2.52). In this equation, the Fock operator $\hat{F}_{rs}$ is given as

$$\hat{F}_{rs} = \left(\sum_{I,J}^{n_{\text{active CSF}}} C_I^* C_J A_{sr}^{IJ}\right)\hat{h} + \sum_{t,u}^{n_{\text{orb}}}\left(\sum_{I,J}^{n_{\text{active CSF}}} C_I^* C_J B_{su,rt}^{IJ}\right) V_{tu}^{\text{ee}}, \tag{3.28}$$

where A_{sr}^{IJ}, $B_{su,rt}^{IJ}$, and V_{tu}^{ee} are represented as

$$A_{sr}^{IJ} = \frac{1}{N!}\int d^3\{\mathbf{x}\}\Phi_I^*(\{\mathbf{x}\})\Phi_J^{s\to r}(\{\mathbf{x}\}), \tag{3.29}$$

$$B_{su,rt}^{IJ} = \frac{1}{N!}\int d^3\{\mathbf{x}\}\Phi_I^*(\{\mathbf{x}\})\Phi_J^{sr\to ut}(\{\mathbf{x}\}), \tag{3.30}$$

and

$$V_{tu}^{\text{ee}} = \int d^3\mathbf{r}_2\phi_t^*(\mathbf{r}_2)\frac{1}{r_{12}}\phi_u(\mathbf{r}_2). \tag{3.31}$$

This equation is practically solved as follows: first, the CI Hamiltonian matrix elements in Eq. (3.24) are calculated for an MCSCF wavefunction in Eq. (3.26). Then, the matrix is diagonalized to obtain the CI coefficients. Using the CI coefficients, the Fock operator in Eq. (3.28) is calculated, and then the MCSCF equation in Eq. (3.27) is solved by the SCF method. After constructing the CSFs, using the given orbitals called *natural orbitals*, the CI Hamiltonian matrix in Eq. (3.24) is diagonalized to obtain the CI coefficients. These processes are repeated to achieve self-consistent CI coefficients. In actual MCSCF calculations, the density matrix is used without explicitly determining the natural orbitals. Natural orbitals can be obtained by diagonalizing the density matrix in the CASSCF method mentioned later, although these orbitals cannot be given in general MCSCF methods using specific excited configurations due to the lack of the degrees of freedom in the unitary transformations of the natural orbitals. Although the electron correlation given in MCSCF calculations is frequently interpreted as nondynamical electron correlation, the dynamical type of electron correlation is included to a greater extent than the nondynamical type. Therefore, *nondynamical correlation should be interpreted as the energy difference between a single-reference calculation*, which uses the Hartree–Fock wavefunction as the reference function, *and the corresponding multireference calculation* (see Sect. 3.5), which uses the MCSCF wavefunction as the reference function. This nondynamical correlation, which is also called the *near-degeneracy effect*, comes from the interaction between near-degenerate electron configurations.

The cause of large electron correlations between near-degenerate electron configurations can be recognized by considering the MP2 energy, i.e., second-order Møller–Plesset, in Eq. (3.13),

$$E_{\mathrm{MP2}} = E_{\mathrm{HF}} + \sum_{I}^{n_{\mathrm{D-CSF}}} \frac{|\int d^3\{\mathbf{x}\}\Phi_{\mathrm{HF}}^*(\{\mathbf{x}\})\hat{H}\Phi_I(\{\mathbf{x}\})|^2}{E_{\mathrm{HF}} - E_I}, \tag{3.32}$$

where $n_{\mathrm{D-CSF}}$ is the number of doubly-excited CSFs. In the case in which an excited configuration has an energy close to that of the ground configuration, the denominator of the perturbation term in Eq. (3.32) approaches zero, giving a large second-order perturbation energy. The above-mentioned bond dissociation of the hydrogen molecule is a good example. Since the ground configuration of the bonding orbital comes close to the excited configuration of the anti-bonding orbital, the electronic energy is stabilized by mixing these configurations.

The most famous MCSCF method is the complete active-space (*CAS*) *SCF method* (Roos et al. 1980), which incorporates all possible excited CSFs in the set of specific valence orbitals. Since the CASSCF method may be the simplest way to take into account the nondynamical electron correlation, this method has been applied to a wide variety of systems from small molecules to biomolecules. However, this method still has various problems: e.g., the number of excited CSFs exponentially increases as the size of active space increases, the SCF process is usually converged poorly in comparison with that of the Hartree–Fock calculation, and the electron correlation is ill-balanced due to the insufficient dynamical correlation.

3.4 Brillouin Theorem

By considering configuration interaction, the characteristics of the Hartree–Fock method become clear. The *Brillouin theorem* (Brillouin 1934) establishes that the Hartree–Fock ground electron configuration has no direct interaction with singly-excited electron configurations (McWeeny 1992). Let us consider the nondiagonal term of the CI Hamiltonian matrix, which consists of the ground configuration and the singly-excited configurations from occupied orbital ϕ_i to virtual orbital ϕ_a. Similarly to the diagonal term in Eq. (2.37), this nondiagonal term is written as

$$\langle\Phi_{\mathrm{HF}}|\hat{H}|\Phi_{i\to a}\rangle = \frac{1}{N!}\int d^3\{\mathbf{r}\} \det|\phi_1^*(\mathbf{r}_1)\cdots\phi_i^*(\mathbf{r}_i)\cdots\phi_N^*(\mathbf{r}_N)|\,\hat{H} \times \det|\phi_1(\mathbf{r}_1)\cdots\phi_a(\mathbf{r}_i)\cdots\phi_N(\mathbf{r}_N)|\,. \tag{3.33}$$

Following the derivation to Eqs. (2.42) and (2.43), the one-electron integral in Eq. (3.33) is found to remain

$$\text{(One-electron integral)} = \int d^3\mathbf{r}\phi_i^*(\mathbf{r})\left\{-\frac{1}{2}\nabla^2 + V_{\text{ne}}\right\}\phi_a(\mathbf{r}) = \langle i|h|a\rangle. \tag{3.34}$$

This derivation uses the following assumptions: the one-electron operator itself is independent of the orbitals, this integral is not affected by the exchanges of every orbital pair excluding ϕ_i and ϕ_a, and this integral is zero for exchanges of ϕ_i and other orbitals. Considering that the exchanges of ϕ_i and other orbitals are allowed, the two-electron integral is similarly found to remain

$$\begin{aligned}
&\text{(Two-electron integral)}\\
&= \frac{1}{2}\sum_{j\neq i}^{n}\left\{\int d^3\mathbf{r}_1 d^3\mathbf{r}_2\phi_i^*(\mathbf{r}_1)\phi_j^*(\mathbf{r}_2)\frac{1}{r_{12}}\left[\phi_j(\mathbf{r}_1)\phi_a(\mathbf{r}_2) - \phi_a(\mathbf{r}_1)\phi_j(\mathbf{r}_2)\right]\right.\\
&\quad \left. - \int d^3\mathbf{r}_1 d^3\mathbf{r}_2\phi_j^*(\mathbf{r}_1)\phi_i^*(\mathbf{r}_2)\frac{1}{r_{12}}\left[\phi_a(\mathbf{r}_1)\phi_j(\mathbf{r}_2) - \phi_j(\mathbf{r}_1)\phi_a(\mathbf{r}_2)\right]\right\}\\
&= \sum_{j}^{n}\left[\langle ij|aj\rangle - \langle ij|ja\rangle\right].
\end{aligned} \tag{3.35}$$

Note that the condition of $j \neq i$ is deleted in the sum due to the cancellation for $j = i$. Therefore, the nondiagonal term becomes

$$\langle\Phi_{\text{HF}}|\hat{H}|\Phi_{i\to a}\rangle = \langle i|h|a\rangle + \sum_{j}^{n}\left[\langle ij|aj\rangle - \langle ij|ja\rangle\right]. \tag{3.36}$$

Using Eq. (2.51), it can be shown that this nondiagonal term corresponds to the nondiagonal term of the Fock matrix,

$$\begin{aligned}
\int d^3\mathbf{r}\phi_i^*(\mathbf{r})\hat{F}\phi_a(\mathbf{r}) &= \langle i|h|a\rangle + \sum_{j}^{n}\langle i|2\hat{J}_j - \hat{K}_j|a\rangle\\
&= \langle\Phi_{\text{HF}}|\hat{H}|\Phi_{i\to a}\rangle.
\end{aligned} \tag{3.37}$$

However, this nondiagonal term should be zero in Hartree–Fock SCF calculations, because by multiplying ϕ_a^* to both sides of Eq. (2.52) and integrating them with respect to the electron coordinate, the right-hand side is shown to give zero due to the orthonormalization condition for orbitals. This indicates that the *ground electron configurations given by solving the one-electron SCF equations such as the Hartree–Fock SCF equation imply the configuration interactions with singly-excited* *electron configurations*. Therefore, it is essentially unnecessary to mix singly-excited configurations with the Hartree–Fock wavefunctions. This is called the Brillouin theorem. Note that if doubly-excited electron configurations are taken into consideration, singly-excited electron configurations should be mixed due to their indirect configuration interactions.

3.5 Advanced Correlation Theories

Thus far, several major theories for incorporating electron correlations have been introduced. However, each theory has some problems, e.g., long computational times required and ill-balanced electron correlations included. This has long been one of the most significant subjects, i.e., how to incorporate electron correlations efficiently and elegantly. In this section, several theories developed to address this issue are briefly reviewed.

The *coupled cluster method* is a major theory developed to efficiently accumulate electron correlations (McWeeny 1992). This method was developed by Čížek based on the ansatz of the cluster expansion in statistical mechanics (Čížek 1966) and became used in quantum chemistry calculations in the 1980s. In this method, the wavefunction is expanded as

$$\Psi^{\mathrm{CC}} = \exp(\hat{T})\Phi = \left(1 + \hat{T} + \frac{1}{2}\hat{T}^2 + \cdots\right)\Phi, \tag{3.38}$$

where Φ is the Slater determinant of the ground electron configuration. $\hat{T}$ in Eq. (3.38) is the operator summing all excitations and similarly to Eq. (3.23), it is written as

$$\hat{T}\Phi = \hat{T}_1\Phi + \hat{T}_2\Phi + \cdots \tag{3.39}$$

$$= \sum_{i}^{n_{\mathrm{occ}}}\sum_{a}^{n_{\mathrm{vir}}} c_{i\to a}\Phi_{i\to a} + \frac{1}{4}\sum_{i,j}^{n_{\mathrm{occ}}}\sum_{a,b}^{n_{\mathrm{vir}}} c_{ij\to ab}\Phi_{ij\to ab} + \cdots, \tag{3.40}$$

where $\hat{T}_1$, $\hat{T}_2$, ... are the operators of single, double, ... excitations, respectively. Using these operators, Eq. (3.38) is represented as

$$\Psi^{\mathrm{CC}} = \exp(\hat{T})\Phi = \left(1 + \hat{T}_1 + \hat{T}_2 + \frac{1}{2}\hat{T}_1^2 + \hat{T}_1\hat{T}_2 + \frac{1}{2}\hat{T}_2^2 + \cdots\right)\Phi. \tag{3.41}$$

The coupled cluster method is the CI method based on this wavefunction. One of the main advantages of this method is to ensure size consistency (see Sect. 3.3) in the selection of electron configurations. In coupled cluster methods, the coupled-cluster singles and doubles (*CCSD*) *method* terminates $\hat{T}$ up to the double excitation operator $\hat{T}_2$ ($\hat{T} = \hat{T}_1 + \hat{T}_2$), and the CCSD with perturbative triples (*CCSD(T)) method* supplements the perturbation effect coming from the triple excitation operator to the CCSD operator. The CCSD(T) method has even been called the *golden theory*, because this method is known to give well-balanced electron correlations in practical computational times. The well-balanced electron correlations are interpreted to come from the partial inclusion of nondynamical electron correlations in the perturbation terms of triple excitations besides dynamical electron correlations from single and double excitation operators. The symmetry-adapted cluster (*SAC*) *CI method* is an application of this coupled cluster method to excited states (Nakatsuji and Hirao 1978). This method incorporates electron correlations of excited states based on those of ground states to shorten the computational time required in excited state calculations.

The *multireference theory* is a significant theory for considering the characteristics of electron correlation. In this method, a dynamical correlation is supplemented into an MCSCF method to balance dynamical and nondynamical electron correlations. This method incorporates excited configurations from a multiconfigurational reference wavefunction to estimate configuration interactions. Whitten and Hackmeyer proposed the *multireference CI (MRCI) method* as the first multireference theory (Whitten and Hackmeyer 1969). However, MRCI calculations are impractical even with present-day computational resources (2014) except for calculations on quite small molecules. For shortening the computational time, the most frequently used multireference theories may be the *multireference perturbation methods*: e.g., CAS second-order perturbation theory (*CASPT2*) (Andersson et al. 1990), the multireference Møller–Plesset perturbation (*MRMP*) *method* (Hirao 1992), and multiconfigurational quasi-degenerate perturbation theory (*MCQDPT*) (Nakano 1993). *Multireference methods are well suited especially for excited state calculations, in which well-balanced electron correlations are requisite, and are highly accurate to such an extent that the calculated results are used to establish the validities of other excited state calculations and even experiments.* Note, however, that even these methods are available only for the calculations on relatively small molecules due to their long computational times. On that note, in the CASSCF method giving the reference function, the number of excited CSFs exponentially increases with the number of active valence orbitals, as indicated in Sect. 3.3. To solve this problem, the density matrix renormalization group (*DMRG*) *method*, which was developed to calculate the quantum lattice model in solid state physics (White 1992; White and Martin 1999), has recently been applied to the CASSCF method. This DMRG method contracts the multielectron basis representation, which corresponds to the eigenvector of the density matrix, for basis functions separated by distances greater than a given correlation length by using renormalization group transformations. Applying this method to the CASSCF method drastically reduces the computational times for CASSCF calculations while maintaining their accuracy (Yanai and Chan 2006). This method has recently been applied to multireference perturbation theories (Kurashige and Yanai 2009) to perform highly accurate calculations of large systems, which have never been targetted in multireference calculations.

Finally, as an efficient correlation method, let us examine the *explicitly correlated method* (Klopper et al. 2006), which drastically enhances the convergence to basis set limit. This method is grounded in the concept of the *Hylleraas CI (HCI) method* (Hylleraas 1929), in which electron correlation is efficiently incorporated by adding a term explicitly dependent on the interelectron distance r_{12}. Although this method has been forgotten for many years, Kutzelnigg revived it as a sophisticated method for efficiently incorporating electron correlation (Kutzelnigg 1985). Kutzelnigg and Klopper developed the *linear R12 method*, which improves the basis-set convergence of the CI method based on the wavefunction expansion of the linear r_{12} term (Kutzelnigg 1991). Ten-no proposed the *F12 method*, which further enhances the convergence by replacing this linear r_{12} term with the $\exp(-\zeta r_{12})$ term (Ten-no 2004). These linear R12 and F12 methods have been applied to the above-mentioned coupled cluster method and have succeeded in achieving high convergence of electron correlation to the basis-set limit.

References

Andersson, K., Malmqvist, P.A., Roos, B.O., Sadlej, A.J., Wolinski, K.: J. Phys. Chem. **94**, 5483–5488 (1990)
Brillouin, L.: Actual. Sci. Ind. **71**, 159–160 (1934)
Čížek, J.: J. Chem. Phys. **45**, 4256–4266 (1966)
Condon, E.U.: Phys. Rev. **36**, 1121–1133 (1930)
Frenkel, J.: Wave Mechanics: Advanced General Theory. Clarendon Press, Oxford (1934)
Frye, D., Preiskorn, A., Lie, G.C., Clementi, E.: In: Clementi, E. (ed.) MOTECC Modern Techniques in Computational Chemistry. ESCOM, Leiden (1990)
Hirao, K.: Chem. Phys. Lett. **190**, 374–380 (1992)
Hylleraas, E.A.: Z. Phys. **54**, 347–366 (1929)
Kato, T.: Commun. Pure Appl. Math. **10**, 151–177 (1957)
Klopper, W., Manby, F.R., Ten-no, S., Valeev, E.F.: Int. Rev. Phys. Chem. **25**, 427–468 (2006)
Kurashige, Y., Yanai, T.: J. Chem. Phys. **130**, 234114(1–21) (2009)
Kutzelnigg, W.: Theor. Chem. Acc. **68**, 445–469 (1985)
Kutzelnigg, W.: J. Chem. Phys. **94**, 1985–2001 (1991)
Löwdin, P.-O.: Phys. Rev. **97**, 1509–1520 (1955)
McWeeny, R.: Methods of Molecular Quantum Mechanics, 2nd edn. Academic, San Diego (1992)
Møller, C., Plesset, M.S.: Phys. Rev. **46**, 618–622 (1934)
Nakano, H.: J. Chem. Phys. **99**, 7983–7992 (1993)
Nakatsuji, H., Hirao, K.: J. Chem. Phys. **68**, 2053–2065 (1978)
Pople, J.A., Seeger, R., Krishnan, R.: Int. J. Quantum Chem. **12**, 149–163 (1977)
Roos, B.O., Taylor, P.R., Siegbahn, P.E.M.: Chem. Phys. **48**, 157–173 (1980)
Sinanoğlu, O.: Adv. Chem. Phys. **6**, 315–412 (1964)
Slater, J.C.: Phys. Rev. **34**, 1293–1322 (1929)
Ten-no, S.: Chem. Phys. Lett. **398**, 56–61 (2004)
White, S.R.: Phys. Rev. Lett. **69**, 2863–2866 (1992)
White, S.R., Martin, R.L.: J. Chem. Phys. **110**, 4127–4130 (1999)
Whitten, J.L., Hackmeyer, M.: J. Chem. Phys. **51**, 5584–5596 (1969)
Yanai, T., Chan, G.K.: J. Chem. Phys. **124**, 194106(1–16) (2006)

Chapter 4
Kohn–Sham Method

4.1 Thomas–Fermi Method

In the field of quantum chemistry, theories on electronic motion states commencing with the Hartree–Fock method have intended to incorporate electron correlation more efficiently. However, due to the long computational times required, quantum chemistry calculations had been mostly restricted to trial applications by theoreticians until the 1980s. In the 1990s, density functional theory (DFT) appeared in quantum chemistry to resolve this situation. After this, DFT has become widespread, so that is now the main theory, which is used in more than 80% of quantum chemistry papers in 2014.

The basic concept of DFT is to make it possible to perform high-speed calculations of many-electron systems by representing the potential as the functional not of the orbitals but of the electron density. The name "functional" indicates the function of a function, which is the electron density in this case. This basic concept was suggested in 1927, only one year after the Schrödinger equation was developed and the same year that the Hartree method was proposed. To solve the Schrödinger equation for the electronic motion states of solid crystals, Thomas suggested this basic concept of DFT (Thomas 1927). In his theory, an electronic state of a *uniform electron gas* is assumed to be a solution of the Schrödinger equation based on electron density. That is, for electronic motions in solid crystals, electrons are distributed uniformly at a proportion of two per unit cell, which is taken to be a cube with each edge length equal to Planck's constant in six-dimensional phase space. It is also assumed that the external potential, which corresponds to the nuclear–electron interaction potential (see Sect. 2.1) in the absence of an electromagnetic field, depends only on the distances from the nuclei and is therefore determined by the nuclear charge and electron density. Based on these assumptions, the kinetic energy functional of the electron density ρ in Eq. (2.76) is formulated as

$$T^{\mathrm{TF}} = C_{\mathrm{F}} \int d^3\mathbf{r}\rho^{5/3}(\mathbf{r}), \tag{4.1}$$

T. Tsuneda, *Density Functional Theory in Quantum Chemistry*,
DOI 10.1007/978-4-431-54825-6_4, © Springer Japan 2014

where

$$C_F = \frac{3}{10}(3\pi^2)^{2/3}. \tag{4.2}$$

This kinetic energy functional was the first *local density approximation (LDA)*. In the next year (1928), Fermi independently derived the same kinetic energy functional as Thomas's functional using Fermi statistics at the absolute zero point, completing what is now known as the *Thomas–Fermi method*, on the basis of the Hartree method (see Sect. 2.1) (Fermi 1928).

Although the Thomas–Fermi method is an interesting theory representing the Hamiltonian operator as the functional only of the electron density, even qualitative discussions cannot be contemplated based on this method in actual electronic state calculations. Dirac considered that this problem may be attributed to the lack of exchange energy (see Sect. 2.4), which was proposed in the same year (Fock 1930), and proposed the first *exchange functional* of electron density ρ (Dirac 1930),

$$E_x^{\text{LDA}} = -\frac{3}{4}\left(\frac{3}{\pi}\right)^{1/3} \int d^3\mathbf{r}\rho^{4/3}(\mathbf{r}). \tag{4.3}$$

This functional is found to be the exact LDA exchange functional. Furthermore, von Weizsäcker proposed a correction term using the gradient of electron density for the Thomas–Fermi kinetic energy functional (von Weizsäcker 1935),

$$T^{\text{W}} = \frac{1}{8}\int d^3\mathbf{r}\frac{|\nabla\rho(\mathbf{r})|^2}{\rho(\mathbf{r})}. \tag{4.4}$$

The 1/9 value of this term was later proven to be the exact correction term for the Thomas–Fermi kinetic energy functional (Parr and Yang 1994). Since this correction term contains the gradient of electron density $\nabla\rho$, it is taken as the first *generalized gradient approximation (GGA)*.

Although various attempts have been made to modify the Thomas–Fermi method, all of these attempts have failed to make the method reliable, because it has no physical background establishing the uniqueness of solutions and the existence of density functionals, and it also cannot reproduce even chemical bonds qualitatively. As a consequence, this method had been forgotten until the mid-1960s.

4.2 Hohenberg–Kohn Theorem

In 1964, the concept of the Thomas–Fermi method was revived by a theorem called the *Hohenberg–Kohn theorem* (Hohenberg and Kohn 1964). This theorem consists of the following two subsidiary theorems for nondegenerate ground electronic states:

1. *External potentials*, which correspond to the nuclear–electron interaction potentials in the absence of an electromagnetic field, *are determined by the electron density.*
2. *The energy variational principle is always established for any electron density.*

Since these theorems were also proven mathematically, establishing the validity of the concept (Kutzelnigg 2006), they can be interpreted as the basic theorems of a quantum theory based on electron density.

The first theorem establishes that external potentials can be represented not by the wavefunction (orbitals in the independent electron approximation) but by the electron density. That is, it confirms that *external potentials and consequently the Hamiltonian operator of ground electronic states can be uniquely determined only by the electron density*. This indicates that all of the information for the Hamiltonian operator of a ground electronic state is included in the electron density (Kutzelnigg 2006). Note, however, that this theorem proves neither the existence of universal density functionals for external potentials nor the uniqueness of electron densities corresponding to external potentials (Kutzelnigg 2006). This theorem was proven by a *reductio ad absurdem* argument (Hohenberg and Kohn 1964). The proof assumes that there are two different external potentials corresponding to the same electron density, and then confirms that this assumption is not consistent with the variational principle for a wavefunction. However, this proof remains a severe problem: it implicitly assumes the unproven one-to-one correspondence of the electron density to the wavefunction. It became clear that this problem, which is called the *V-representability problem*, has a serious implication for the concept of DFT, as mentioned later.

The second theorem proves *the variational principle that the Hamiltonian operator represented by the electron density definitely has a solution of (local) minimum energy*. In the proof of this theorem, it is required to assume the establishment both of the first theorem and the variational principle for the wavefunction. That is, if there is an exact energy functional with an external potential, it can be proven that the electron density is uniquely determined to give a (local) minimum energy for this external potential on the basis of this assumption. The proof of this theorem also still contains two problems. One problem is implicitly assuming the *N-representability* of the electron density. The N-representability of the electron density indicates the following: the electron density is always greater than or equal to zero, the sum of the electron density is the number of electrons N, and the square of the gradient of the square root of the electron density $|\nabla\rho^{1/2}|^2$ has a finite total sum. However, this N-representability of the electron density had not been proven. In general, the energy variation principle is violated unless the Hamiltonian operator is written in terms of N-representable variables. This problem is called the *N-representability problem*. It has already been proven that the electron density in Eq. (2.76) is an N-representable variable (Gilbert 1975). Furthermore, as mentioned in the next section, since DFT calculations are mostly based not on an N-electron wavefunction but on a Slater determinant, which is obviously N-representable, it is rare to discuss the N-representability problem of variables. Instead, it is often seen as a problem that the variational principle is generally not assured in DFT calculations due to approximate density functionals being used, which are usually not based on a Slater determinant (Kutzelnigg 2006).

The above-mentioned V-representability problem is also solved. This problem indicates that the one-to-one correspondence between electron density and wavefunction is essentially not proven. Naturally, the wavefunction has a one-to-one correspondence to the Hamiltonian operator of the ground electronic state. It is also trivial that the wavefunction uniquely determines the electron density. However, the wavefunction is generally not determined uniquely by the electron density. The first theorem is not established without solving this problem. To solve this problem, Levy suggested the *constrained search formulation* (Levy 1979) based on the variational principle for energy. The Hamiltonian operator contains the kinetic energy operator, $\hat{T}$, and the electron–electron interaction operator, $\hat{V}_{ee}$, in addition to the external potential. Therefore, for determining the external potential uniquely from the electron density, it is required only to find the wavefunction giving the lowest energy expectation value of the other two operators for the electron density. That is, it requires the existence of a *universal functional*, E_{univ}, which is represented as

$$E_{univ}[\rho] = \min_{\Psi \to \rho} \int d^3\mathbf{r}\Psi^* \left(\hat{T} + \hat{V}_{ee}\right) \Psi. \tag{4.5}$$

Conversely, the existence of the universal functional establishes the one-to-one correspondence of the electron density with the wavefunction, which is searched to give the lowest energy expectation value for the sum of the kinetic energy and electron–electron interaction operators. The "constrained" designation represents a constraint for the wavefunction, Ψ, to that giving a specific electron density, ρ. For the universal functional, the existence of the lowest energy was proven by Lieb (Lieb 1983). Note that this search removes the restriction to "nondegenerate" electronic states for the Hohenberg–Kohn theorem. Therefore, this constrained search formulation confirms the one-to-one correspondence between the electron density and the wavefunction for every system, and consequently, it solves the V-representability problem.

The constrained search formulation has important implications in DFT. This requires the energy density functional, which gives the lowest energy expectation value for the sum of the kinetic energy and electron–electron interaction operators. Note that this energy density functional is independent of the external potential. This indicates that the universal energy density functionals of the kinetic and electron–electron interaction operators are required, independent of the nuclear coordinates, i.e., target systems. Therefore, *the remaining subject in DFT is to develop universal density functionals giving the lowest energy expectation value for the sum of the kinetic and electron–electron interaction operators.* Since the development of density functionals has actually been the main subject in the field of DFT since the 1980s, it is no exaggeration to say that the idea of the constrained search formulation dictates the subsequent policy of DFT studies.

4.3 Kohn–Sham Method

Although the Hohenberg–Kohn theorem is established as the fundamental theorem of quantum chemistry based on electron density, it is not in itself sufficient to calculate actual electronic motion states. In the next year, Kohn and Sham developed an electronic state calculation method derived from this theorem, which is called the *Kohn–Sham method* (Kohn and Sham 1965). The Kohn–Sham method is a variational approach using the electron–electron interaction potential of the density functional to give the lowest energy and the corresponding molecular orbitals and orbital energies. From a practical viewpoint, the most significant characteristic of this method is *the use of the independent-electron approximation of kinetic energy, similarly to the Hartree–Fock method, instead of the kinetic energy functional in the Thomas–Fermi method.* This simple replacement solved the weakest point of the Thomas–Fermi method, as mentioned in Sect. 4.1. Later, the Kohn–Sham method, using the independent-electron kinetic energy, made it possible to carry out high-speed quantitative electronic state calculations in chemistry and solid state physics through the development of sophisticated exchange-correlation functionals. This has led to an explosion in the use of DFT. Note, however, that, the Kohn–Sham method is not a pure DFT in the spirit of the Thomas–Fermi method. Nevertheless, it should be emphasized that *the Kohn–Sham method is an exact formulation of the Hohenberg–Kohn theorem* (Eschrig 2003) in the sense that electronic structures are determined using the one-to-one correspondence between external potential and electron density on the basis of the variational principle.

In practical calculations making use of the Kohn–Sham method, the *Kohn–Sham equation* is used. This equation is a one-electron SCF equation applying the Slater determinant to the wavefunction of the Hartree method, similarly to the Hartree–Fock method. Therefore, in the same manner as the Hartree–Fock equation, this equation is derived to determine the lowest energy by means of the Lagrange multiplier method, subject to the normalization of the wavefunction (Parr and Yang 1994). As a consequence, it gives a similar Fock operator for the nonlinear equation,

$$\hat{F} = \hat{h} + 2\sum_{j}^{n} \hat{J}_j + V_{\mathrm{xc}}. \tag{4.6}$$

The difference between this Fock operator and the Hartree–Fock counterpart in Eq. (2.51) is only the exchange-correlation potential functional, V_{xc}, which substitutes for the exchange operator in the Hartree–Fock operator. That is, in the electron–electron interaction potential, only the exchange operator is replaced with the approximate potential density functionals of the exchange interactions and electron correlations, while the remaining Coulomb operator, $\hat{J}_j$, which is represented as the interaction of electron densities, is used as is. The point is that the *electron correlations, which are incorporated as the interactions between electron configurations in wavefunction theories* (see Sect. 3.3), *are simply included as a potential density functional.* Using this Fock operator, orbitals, ϕ_i, and orbital energies, ϵ_i are calculated by solving the one-electron equation,

$$\hat{F}\phi_i = \epsilon_i\phi_i, \tag{4.7}$$

similarly to the Hartree–Fock equation. The orbital energies are represented as

$$\begin{aligned}\epsilon_i &= \int d^3\mathbf{r}_1 \phi_i^*(\mathbf{r}_1)\hat{F}\phi_i(\mathbf{r}_1) \\ &= h_i + 2\sum_j^n J_{ij} + \int d^3\mathbf{r}_1 \rho_i(\mathbf{r}_1)V_{\mathrm{xc}},\end{aligned} \tag{4.8}$$

where ρ_i is the electron density of the i-th orbital and is summed to be the total electron density,

$$\rho = \sum_i^n \rho_i = 2\sum_i^n |\phi_i|^2. \tag{4.9}$$

Differently to the Hartree–Fock method, the total electronic energy in the Kohn–Sham method is generally calculated using not the exchange-correlation potential functional but the exchange-correlation energy functional, E_{xc}, as

$$E = \sum_i^n \left(h_i + 2\sum_j^n J_{ij} \right) + E_{\mathrm{xc}}. \tag{4.10}$$

The exchange-correlation potential functional, V_{xc}, is the first derivative of this exchange-correlation energy functional with respect to electron density,

$$V_{\mathrm{xc}} = \frac{\delta E_{\mathrm{xc}}}{\delta\rho}. \tag{4.11}$$

Similarly to the Hartree–Fock method, the Kohn–Sham method is solved by the SCF method for solving nonlinear equations through the following steps:

1. Set up the information on the calculated molecular system (nuclear coordinates, nuclear charges, and number of electrons) and initial reference molecular orbitals.
2. Calculate the two-electron operators of the Fock operator in Eq. (4.6) using the reference orbitals.
3. Solve Eq. (4.7) with the calculated Fock operator.
4. Calculate the exchange-correlation energy using the given orbitals, and then the total electronic energy in Eq. (4.10).
5. Compare the given molecular orbitals and total energy with the previous ones, and take them as the solution if the differences are less than preset thresholds. If the differences are greater than the thresholds, return to process 2 with the given orbitals as the initial orbitals.

Note that *the difference of the Kohn–Sham SCF method from the Hartree–Fock one is only in the points that the total electronic energy is separately calculated in process 4 and it is used as a determining factor in process 5*. This difference is attributed to the problem that the total electronic energy in the Kohn–Sham method is generally not represented using orbital energies. This is because, in the Kohn–Sham method that uses approximate exchange-correlation potential functionals, the expectation value of the exchange-correlation potential is not identical to the exchange-correlation energy, i.e.,

$$E_{\mathrm{xc}} \neq \sum_{i}^{n} \int d^3\mathbf{r}_1 \phi_i^*(\mathbf{r}_1) V_{\mathrm{xc}} \phi_i(\mathbf{r}_1) = \int d^3\mathbf{r}_1 \rho(\mathbf{r}_1) V_{\mathrm{xc}}. \tag{4.12}$$

Therefore, as shown in Eq. (4.10), the total electronic energy in the Kohn–Sham method is separately calculated using the exchange-correlation energy functionals. Note, however, that no such difference should be included in a universal functional, and there is a possibility in the future to develop an exchange-correlation functional giving no difference. In this case, process 4 is not necessary.

The Kohn–Sham equation is also transformed into a matrix equation on the basis of the Roothaan method in Sect. 2.5. Similar to the Hartree–Fock equation, the Kohn–Sham–Roothaan equation is written as

$$\mathbf{F}\mathbf{C}_i = \epsilon_i \mathbf{S}\mathbf{C}_i. \tag{4.13}$$

The elements of matrix **F** are

$$F_{pq} = h_{pq} + \sum_{r,s=1}^{n_{\mathrm{basis}}} P_{rs} \langle pr|qs \rangle + (V_{\mathrm{xc}})_{pq}, \tag{4.14}$$

where

$$(V_{\mathrm{xc}})_{pq} = \int d^3\mathbf{r} \chi_p^*(\mathbf{r}) V_{\mathrm{xc}} \chi_q(\mathbf{r}). \tag{4.15}$$

Terms other than the exchange-correlation potential are the same as those in Sect. 2.5. In this method, the total electronic energy is separately calculated as different to that in the Roothaan method.

4.4 Generalized Kohn–Sham Method

Although the Kohn–Sham method has been the basic procedure in DFT calculations, many exchange-correlation functionals used in conventional DFT calculations have no strict theoretical basis upon which to be used in the Kohn–Sham method. Since the Kohn–Sham method is based on the constrained search formulation, it is proven to be applicable to pure exchange-correlation energy density functionals, which are formed only from the electron density. However, non-pure functionals such as hybrid functionals (see Sect. 5.5) and long-range corrected functionals (see Sect. 6.1) are not appropriate in the original Kohn–Sham method. For example, it is not strictly correct to use the most widely used B3LYP functional within the framework of the Kohn–Sham method, because it contains the Hartree–Fock exchange integral, which is not formed from the electron density. To solve this problem, Levy and coworkers extended the constrained search formulation to that using functionals containing the Hartree–Fock exchange integral, and suggested the *generalized Kohn–Sham method* as an extension of the Kohn–Sham method (Seidl et al. 1986). In this method, the action, S, is defined as a functional summing terms given by not the N-electron wavefunction but the Slater determinant, Φ: the kinetic and Coulomb energies with the Hartree–Fock exchange integral term, which is partly included in hybrid and long-range corrected functionals, such as

$$S[\Phi] = \int d^3\mathbf{r}\Phi^*\hat{T}\Phi + \sum_{i,j}^{n}\left(2J_{ij}(\{\phi_i\}) - K_{ij}^{\text{HF-part}}(\{\phi_i\})\right). \tag{4.16}$$

The first Hohenberg–Kohn theorem for this method is established if there is a universal functional, F^S, that uniquely determines the electron density, giving the least action:

$$F^S[\rho] = \min_{\Phi\to\rho(\mathbf{r})} S[\Phi] = \min_{\{\phi_i\}\to\rho(\mathbf{r})} S[\{\phi_i\}]. \tag{4.17}$$

Using this S, the total electronic energy is defined as

$$E^S[\{\phi_i\}; V_{\text{eff}}] = S[\{\phi_i\}] + \int d^3\mathbf{r}\rho(\mathbf{r})V_{\text{eff}}[\rho(\mathbf{r})], \tag{4.18}$$

and the corresponding one-electron equation is given as

$$\hat{O}^S[\{\phi_i\}]\phi_j + V_{\text{eff}}\phi_j = \epsilon_j\phi_j \quad (j = 1, \cdots, N), \tag{4.19}$$

where $\hat{O}^S$ is the orbital-dependent operator in the Fock operator and V_{eff} is the effective potential functional for the remaining orbital-independent part, which is usually the density functional part of the exchange-correlation potential functionals. The advantages of this method are *the wide applicabilities to density functionals, which accept various types of S functionals, and the minimal effect of the density functional part on the electron density*. In particular, the latter reduces the importance of the variational principle problem resulting from the use of approximate functionals, as mentioned in the previous section. On the other hand, this method produces another problem: *this method assumes the existence of a universal functional, F^S, which gives a one-to-one correspondence between the Slater determinant and the electron density*. Actually, the S functional is invariant for the unitary transformation of orbitals, because it is formed from the Slater determinant. Since the unitary transformation of orbitals changes the electron density, it is difficult to uniquely determine the electron density that minimizes S. Nevertheless, the remaining effective potential functional of the electron density, V_{eff}, has no degrees of freedom for the unitary transformation to uniquely determine the electron density. This functional, consequently, has a one-to-one correspondence with the electron density. Therefore, the generalized Kohn–Sham method is a theory that formulates the Hohenberg–Kohn theorem more comprehensively than the Kohn–Sham method, which targets only pure electron density functionals. Henceforth, "the Kohn–Sham method" in which functions other than electron density can be mixed in the functionals, is assumed, in this book, to be within the framework of the generalized Kohn–Sham method, without explicitly including the "generalized" designation.

4.5 Constrained Search Method for Constructing Kohn–Sham Potentials

The one-to-one correspondence between electron density and effective potential, which is proven on the basis of the constrained search formulation, suggests that *the effective potential can be determined directly from the electron density*. Parr and coworkers developed a procedure for determining highly accurate exchange-correlation potentials from electron densities, which are calculated by high-level *ab initio* correlation wavefunction theories. This procedure is called the *Zhao–Morrison–Parr (ZMP) method* (Zhao et al. 1994). In this method, the effective potential is given by the Lagrange undetermined multiplier method with a potential,

$$V_{\mathrm{constr}}^{\lambda}(\mathbf{r}) = \lambda \int \frac{\rho(\mathbf{r}') - \rho_0(\mathbf{r})}{|\mathbf{r} - \mathbf{r}'|} d^3\mathbf{r}', \tag{4.20}$$

which is zero for the given electron density, as the constraint, such as

$$\left[\hat{h} + 2\left(1 - \frac{1}{n}\right)\sum_{j}^{n} \hat{J}_j^{\lambda} + V_{\mathrm{constr}}^{\lambda}(\mathbf{r})\right]\phi_i^{\lambda} = \epsilon_i^{\lambda}\phi_i^{\lambda}, \tag{4.21}$$

where $\hat{J}_j^{\lambda}$ indicates the Coulomb potential, $\hat{J}_j$, in which the electron density corresponding to each λ is employed. This equation is solved by an iterative approach to provide the effective potential at $\lambda = \infty$, as mentioned later. At $\lambda = \infty$, this equation becomes the Kohn–Sham equation. Note that the $(1 - 1/n)$ term is multiplied by the Coulomb operator. This term comes from the Fermi–Amaldi self-interaction correction (Fermi and Amaldi 1934) (see Sect. 6.2). By use of this term, the given effective potential can be interpreted as the exchange-correlation potential,

$$V_{\mathrm{xc}} = \lim_{\lambda \to \infty}\left[V_{\mathrm{constr}}^{\lambda}(\mathbf{r}) - \frac{2}{n}\sum_{j}^{n} \hat{J}_j^{\lambda}\right]. \tag{4.22}$$

This also makes a finite λ available and upgrades the convergence and accuracy of the SCF method.

The ZMP method is carried out according to the following procedure:

1. First, calculate the terms in Eq. (4.21) using the exact electron density, ρ_0.
2. Setting an appropriate λ, solve Eq. (4.21) by an SCF calculation.
3. If λ does not reach the preset limit value, set a larger λ and then return to process 2 with the given orbitals and orbital energies. If λ reaches the preset limit value, the calculated orbitals and orbital energies correspond to the exact electron density.
4. Using the calculated orbitals and orbital energies, determine the exchange-correlation potential in Eq. (4.22).

Since it is generally difficult to obtain the exact electron density in actual calculations, this method uses the calculated electron densities of advanced correlation wavefunction theories such as multireference theories (see Sect. 3.5). Consequently, Parr et al. succeeded in obtaining a highly accurate exchange-correlation potential, and found that *the exchange-correlation potential corresponding to the Kohn–Sham method is neither uniform nor local.* This result justifies the use of functionals combining nonuniform nonlocal potentials such as the Hartree–Fock exchange potential with (semi-)local exchange-correlation potential functionals.

One of the most significant findings to which the ZMP method contributes is the comprehension of the reasons why the Kohn–Sham method can accurately reproduce chemical behavior. Baerends and coworkers compared highly accurate exchange-correlation energies, which are calculated by multiplying the ZMP potential of the MRCI electron density (for MRCI, see Sect. 3.5) by the electron density, with exchange-correlation energy functionals (Fig. 4.1) (Schipper et al. 1998). The figure shows that, even though correlation functionals are much different from the correlation energy density of the MRCI calculation, exchange-correlation functionals behave similarly to the MRCI exchange-correlation energy density. Actually, conventional correlation functionals usually contain only dynamical electron correlation, stemming from the short-range interelectron correlation cusp (see Sect. 5.3). In Sect. 3.2, it is explained that this electron correlation is classified into dynamical and nondynamical electron correlations. This figure reveals that *exchange-correlation functionals include dynamical correlations in correlation functionals and nondynamical correlations in exchange functionals, and consequently provide accurate energy densities, similar to those of MRCI as a whole.* Reproducing chemical properties and chemical reactions quantitatively requires incorporating well-balanced electron correlations. Therefore, this leads to a clear indication of a reason for the high reproducibility of the Kohn–Sham method in calculations of chemical phenomena.

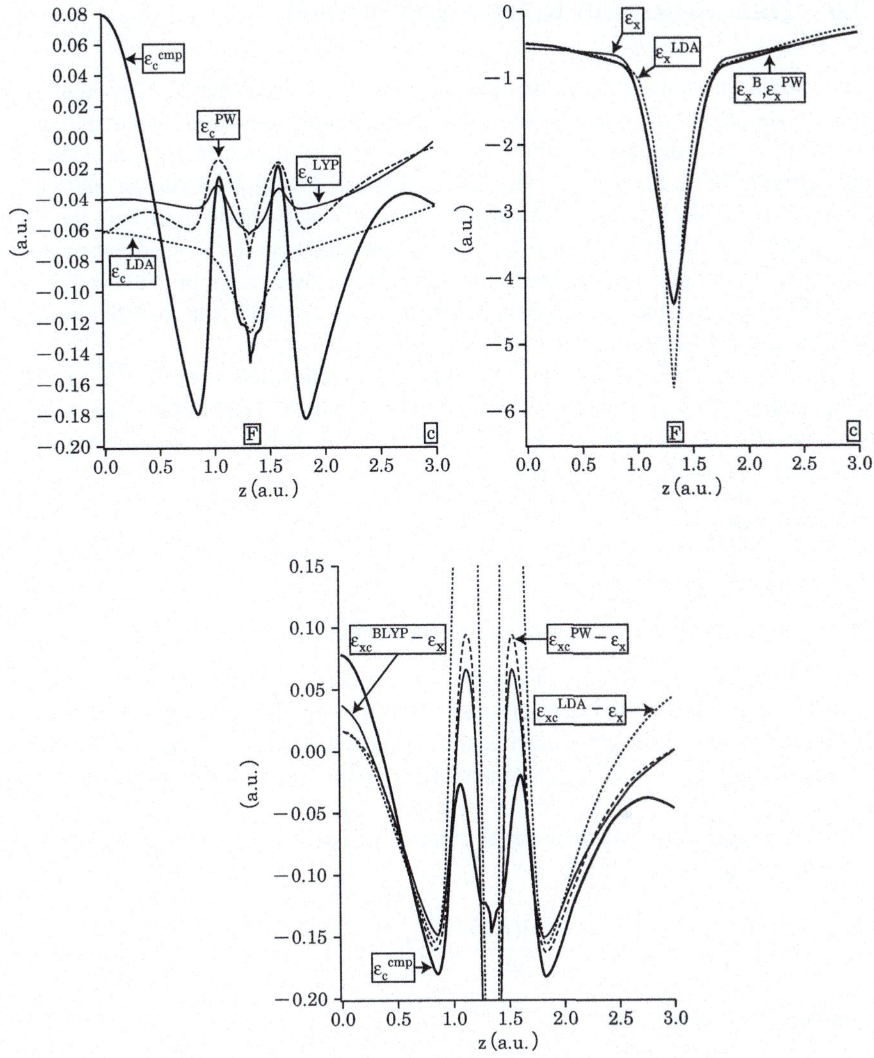

Fig. 4.1 The highly accurate exchange, correlation and exchange-correlation energy densities, ϵ_x, ϵ_c, and ϵ_{xc}, of the F_2 molecule (emp), which are determined by the ZMP method from the MRCI electron density, and the corresponding GGA energy functionals (LDA, Becke 1988 (B) and Perdew–Wang (PW) exchange with the LDA and Lee–Yang–Parr (LYP) correlation functionals mentioned in Chap. 5) for correlation (*upper left*), exchange (*upper right*), and exchange-correlation (*lower*) energies (Schipper et al. 1998)

4.6 Time-Dependent Kohn–Sham Method

Thus far, discussions of the Kohn–Sham method have targeted stationary electronic motion states, i.e., time-independent electronic states. However, actual studies of chemical phenomena need to analyze the time propagation of molecular electronic states induced by nuclear motion, light irradiation, and so forth. In particular, chemists have recently been drawing attention to observing or controling the time propagation of photochemical reactions as observed in time-resolved experiments using ultra-short pulsed-laser beams. It is, therefore, significant, in order to increase its usefulness, to make the Kohn–Sham method available for investigations of time-dependent electronic states.

For this purpose, Runge and Gross proposed *the fundamental DFT theorem for periodically time-dependent electronic states*, which is called the *Runge–Gross theorem* (Runge and Gross 1984). This theorem is based on the following two assumptions for the external potential, V_{ext}:

1. $V_{\text{ext}}(\mathbf{r}, t)$ depends on time periodically.
2. $V_{\text{ext}}(\mathbf{r}, t)$ consists of a time-independent stationary part, $V_{\text{ext}}^{\text{stat}}$, and a slightly time-dependent perturbation part, $V_{\text{ext}}^{\text{pert}}$.

Under these assumptions, the following four theorems are derived:

1. *The first time-dependent Hohenberg–Kohn theorem.* For V_{ext} expansible in terms of time, define that the $V_{\text{ext}}(\mathbf{r}, t) \rightarrow \rho(\mathbf{r}, t)$ transformation corresponds to solving the time-dependent Schödinger equation. Based on this definition, the inverse transformation, $\rho \rightarrow V_{\text{ext}}$, can be performed in the case of the second assumption above.
2. The time-derivative of the current density, $\mathbf{j}$, (see Sect. 6.5) is represented as a density functional, $\Omega[\rho](\mathbf{r}, t)$,

$$\frac{\partial \mathbf{j}(\mathbf{r}, t)}{\partial t} = \Omega[\rho](\mathbf{r}, t), \tag{4.23}$$

where $\mathbf{j}$ is defined as

$$\frac{\partial \rho(\mathbf{r}, t)}{\partial t} = -\nabla \cdot \mathbf{j}(\mathbf{r}, t). \tag{4.24}$$

3. *The second time-dependent Hohenberg–Kohn theorem.* The action integral in an arbitrary time interval from t_0 to t_1,

$$S = \int_{t_0}^{t_1} dt \Psi^*[\rho] \left(i\frac{\partial}{\partial t} - \hat{H} \right) \Psi[\rho], \tag{4.25}$$

is representable as a density functional, $S[\rho]$, and is decomposable to

$$S[\rho] = \int_{t_0}^{t_1} dt \Psi^*[\rho]\left[i\frac{\partial}{\partial t} - \left(\hat{T} + \hat{V}_{ee}\right)\right]\Psi[\rho] - \int_{t_0}^{t_1} dt \int d^3\mathbf{r}\rho(\mathbf{r},t)V_{ext}(\mathbf{r},t), \quad (4.26)$$

where $\Psi[\rho]$ is a wavefunction giving the electron density, ρ. The first term on the right-hand side is a universal functional, because it does not depend on the external potential, V_{ext}. This action density functional, $S[\rho]$, satisfies the variational principle to give the stationary value at the exact density.

4. Time-dependent orbitals, $\{\phi_i(\mathbf{r},t)\}$, satisfy the time-dependent Schödinger equation,

$$i\frac{\partial}{\partial t}\phi_i(\mathbf{r},t) = \left(-\frac{1}{2}\nabla^2 + V_{eff}[\mathbf{r},t;\rho(\mathbf{r},t)]\right)\phi_i(\mathbf{r},t), \quad (4.27)$$

where the effective potential, V_{eff}, is represented as

$$V_{eff}[\mathbf{r},t;\rho(\mathbf{r},t)] = V_{ext}[\mathbf{r},t] + \int d^3\mathbf{r}\frac{\rho(\mathbf{r}',t)}{|\mathbf{r}-\mathbf{r}'|} - \frac{\delta S_{xc}[\rho]}{\delta\rho(\mathbf{r},t)}. \quad (4.28)$$

Note that S_{xc} is the exchange-correlation part of the action integral and is approximated on the basis of the adiabatic approximation (see Sect. 2.2) as

$$\frac{\delta S_{xc}[\rho]}{\delta\rho(\mathbf{r},t)} \approx V_{xc}[\rho](\mathbf{r},t) = \frac{\delta E_{xc}[\rho]}{\delta\rho(\mathbf{r},t)}. \quad (4.29)$$

Equation (4.27) using the effective potential in Eq. (4.28) is called the *time-dependent Kohn–Sham equation* (Runge and Gross 1984).

The Runge–Gross theorem has a severe problem in the use of the wavefunction giving electron density in the third theorem (Gross et al. 1995): Since time-dependent wavefunctions contain time-dependent phase terms, the one-to-one correspondence of the wavefunction and electron density is established only for a specific phase. This problem can be avoided by representing the wavefunction as a functional of the external potential, $\Psi[V_{ext}]$, for which the one-to-one correspondence with electron density is established. However, this approach leads to the denial of the concept of the universal functional. Even if it would be accepted, the one-to-one correspondence of the time-derivative of the wavefunction, $\delta\Psi$ and that of the potential, δV_{ext}, is not assured. Therefore, differently to the time-independent case, *the variational principle is not strictly established in the time-dependent case* (van Leeuwen 2006). Even though one-to-one correspondence can be proven to exist if the external potential is Taylor-expandable in time (Ullrich 2012), it can also be proven that no such Taylor-expandable potential exists (Yang and Burke 2013). This dilemma has been resolved by taking relativistic effects into account, because the

Rajagopal–Callaway theorem, which is the relativistic expansion of the Hohenberg–Kohn theorem, establishes the time-dependent case of the latter (see Sect. 6.4).

By applying the time-dependent Kohn–Sham equation to linear response theory, excitation energies can be calculated and assigned to the corresponding transitions (Gross and Burke 2006). Following the Runge–Gross theorem, assume that only a weak perturbation, δV_{ext}, is added to the external potential. Under this assumption, it is interpreted that the electron density also undergoes an infinitesimal change, $\delta\rho(\mathbf{r},t)$, in the stationary part, ρ_{stat}. The exchange-correlation potential, V_{xc}, is, therefore, represented as

$$\begin{aligned} V_{\mathrm{xc}}[\rho](\mathbf{r}_1,t_1) =& V_{\mathrm{xc}}^{\mathrm{stat}}[\rho](\mathbf{r}_1) \\ &+ \iint dt_2 d^3\mathbf{r}_2 f_{\mathrm{xc}}[\rho_{\mathrm{stat}}](\mathbf{r}_1,\mathbf{r}_2,t_2-t_1)\delta\rho(\mathbf{r}_2,t_2) \end{aligned} \quad (4.30)$$

$$f_{\mathrm{xc}}[\rho_{\mathrm{stat}}](\mathbf{r}_1,\mathbf{r}_2,t_2-t_1) = \left.\frac{\delta V_{\mathrm{xc}}(\mathbf{r}_1,t_1)}{\delta\rho(\mathbf{r}_2,t_2)}\right|_{\rho=\rho_{\mathrm{stat}}}. \quad (4.31)$$

The derivative of the exchange-correlation potential in terms of electron density, f_{xc}, is called the *exchange-correlation integral kernel.* Define the response function of the electron density, χ_{KS}, for the infinitesimal change in the Kohn–Sham potential, δV_{KS}, as

$$\delta\rho(\mathbf{r}_1,t_1) = \iint dt_2 d^3\mathbf{r}_2 \chi_{\mathrm{KS}}[\rho_{\mathrm{stat}}](\mathbf{r}_1,\mathbf{r}_2,t_2-t_1)\delta V_{\mathrm{KS}}(\mathbf{r}_2,t_2). \quad (4.32)$$

In this definition, the response function is given by Green function theory as

$$\begin{aligned} \chi_{\mathrm{KS}}(\mathbf{r}_1,\mathbf{r}_2,\omega) = 2 \lim_{\eta\to 0+} \sum_i^{n_{\mathrm{occ}}} \sum_a^{n_{\mathrm{vir}}} \Bigg[& \frac{\phi_i^*(\mathbf{r}_1)\phi_a(\mathbf{r}_1)\phi_i(\mathbf{r}_2)\phi_a^*(\mathbf{r}_2)}{\omega - (\epsilon_a - \epsilon_i) + i\eta} \\ & - \frac{\phi_i(\mathbf{r}_1)\phi_a^*(\mathbf{r}_1)\phi_i^*(\mathbf{r}_2)\phi_a(\mathbf{r}_2)}{\omega + (\epsilon_a - \epsilon_i) - i\eta} \Bigg]. \end{aligned} \quad (4.33)$$

Note that this response function is Fourier-transformed ($t \to \omega$). What is important is that this *response function has poles in the excitation energies* (see Fig. 4.2). Casida proposed that the pole energies of the response function in Eq. (4.33), i.e., excitation energies, can be calculated by solving the following simultaneous matrix equations (Casida 1996):

$$\sum_{jb\tau} \left[\delta_{\sigma\tau}\delta_{ij}\delta_{ab}\left(\epsilon_{a\sigma} - \epsilon_{i\sigma} + \omega\right) + K_{ia,jb}^{\sigma\tau}\right] X_{jb}^{\tau} + K_{ia,bj}^{\sigma\tau} X_{bj}^{\tau} = 0, \quad (4.34)$$

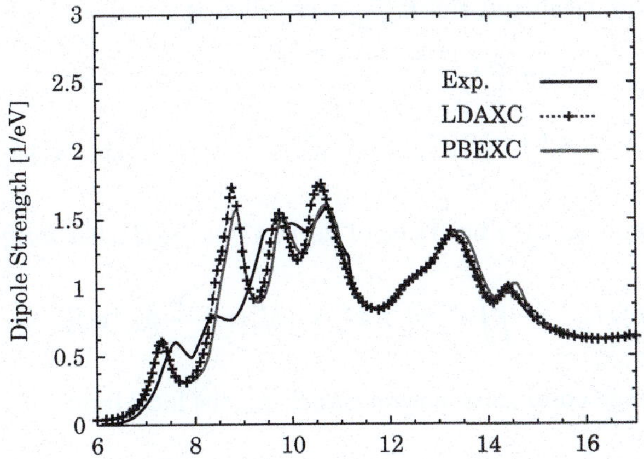

Fig. 4.2 Excitation spectra of the ethane molecule, which were calculated directly from the Kohn–Sham response functions, χ_{KS}, of LDA and PBE exchange-correlation functionals, compared to the experimental one (Exp.) (Marques et al. 2001)

and

$$\sum_{jb\tau}\left[\delta_{\sigma\tau}\delta_{ij}\delta_{ab}\left(\epsilon_{a\sigma}-\epsilon_{i\sigma}+\omega\right)+K_{ai,bj}^{\sigma\tau}\right]X_{bj}^{\tau}+K_{ai,jb}^{\sigma\tau}X_{jb}^{\tau}=0. \quad (4.35)$$

In these equations, the spins of the orbitals (σ, τ, $\sigma' \neq \sigma$) are explicitly displayed for purposes of accuracy. The simultaneous matrix equations are also represented for singlet and triplet excitations as follows:

$$\boldsymbol{\Omega}\mathbf{F}_{ia\sigma} = \omega_{ia}^{2}\mathbf{F}_{ia\sigma}, \quad (4.36)$$

$$\begin{aligned}\Omega_{ia\sigma,jb\tau}^{\mathrm{singlet}} &= \delta_{\sigma\tau}\delta_{ij}\delta_{ab}\left(\epsilon_{a\sigma}-\epsilon_{i\sigma}\right)^{2}\\ &\quad +2\left(\epsilon_{a\sigma}-\epsilon_{i\sigma}\right)^{1/2}\left(K_{ia,jb}^{\sigma\sigma}+K_{ia,jb}^{\sigma\sigma'}\right)\left(\epsilon_{b\tau}-\epsilon_{j\tau}\right)^{1/2}, \end{aligned} \quad (4.37)$$

and

$$\begin{aligned}\Omega_{ia\sigma,jb\tau}^{\mathrm{triplet}} &= \delta_{\sigma\tau}\delta_{ij}\delta_{ab}\left(\epsilon_{a\tau}-\epsilon_{i\tau}\right)^{2}\\ &\quad +2\left(\epsilon_{a\sigma}-\epsilon_{i\sigma}\right)^{1/2}\left(K_{ia,jb}^{\sigma\sigma}-K_{ia,jb}^{\sigma\sigma'}\right)\left(\epsilon_{b\tau}-\epsilon_{j\tau}\right)^{1/2}, \end{aligned} \quad (4.38)$$

where $\mathbf{F}_{ia\sigma}$ is the response coefficient matrix given by

$$F_{ia\sigma} = \left(\epsilon_{a\sigma}-\epsilon_{i\sigma}\right)^{-1/2}\left(X_{ia\sigma}-X_{ai\sigma}\right), \quad (4.39)$$

$$\begin{aligned}X_{ia\sigma}(\omega) &= \frac{-1}{\omega+\left(\epsilon_{a\sigma}-\epsilon_{i\sigma}\right)}\\ &\quad \times\int d^{3}\mathbf{r}\phi_{i\sigma}^{*}(\mathbf{r})\delta\left(2\sum_{i}^{n}\hat{J}_{i}+V_{\mathrm{xc}}\right)(\mathbf{r},\omega)\phi_{a\sigma}(\mathbf{r}), \end{aligned} \quad (4.40)$$

ϵ_i is the i-th orbital energy, and $K_{ia,jb}^{\sigma\tau}$ is provided as

$$K_{ia,jb}^{\sigma\tau} = \langle ib|aj\rangle^{\sigma\tau} + \iint d^3\mathbf{r}_1 d^3\mathbf{r}_2 \phi_{i\sigma}^*(\mathbf{r}_1)\phi_{b\tau}^*(\mathbf{r}_2) f_{\rm xc}(\mathbf{r}_1,\mathbf{r}_2)\phi_{a\sigma}(\mathbf{r}_1)\phi_{j\tau}(\mathbf{r}_2). \quad (4.41)$$

The first term on the right-hand side of Eq. (4.41) is the *Hartree integral*, written as

$$\langle ib|aj\rangle^{\sigma\tau} = \iint d^3\mathbf{r} d^3\mathbf{r}' \phi_{i\sigma}^*(\mathbf{r}_1)\phi_{b\tau}^*(\mathbf{r}_2)\frac{1}{r_{12}}\phi_{a\sigma}(\mathbf{r}_1)\phi_{j\tau}(\mathbf{r}_2). \quad (4.42)$$

For the exchange-correlation integral kernel, $f_{\rm xc}$, the local form,

$$f_{\rm xc}(\mathbf{r}_1,\mathbf{r}_2) = f_{\rm xc}(\mathbf{r}_1)\delta(\mathbf{r}_1 - \mathbf{r}_2), \quad (4.43)$$

is used. Solving Eq. (4.36) gives the sets of excitation energies $\{\omega_{ia}\}$ and corresponding response coefficient matrices, $\mathbf{F}_{ia\sigma}$. Equation (4.36) is called the *time-dependent response Kohn–Sham equation* and the method using this equation is called the *time-dependent response Kohn–Sham method.* In excitation energy calculations, the "time-dependent Kohn–Sham (TDKS) method" usually indicates this method. On the other hand, there is a *time-dependent propagator Kohn–Sham method*, which does not use the linear response theory but instead uses Eq. (4.27) for analyzing the time propagation of electronic states. Note that excitation energies can also be calculated using this method (Yabana and Bertsch 1996). What we should notice is that *the time-dependent response Kohn–Sham method is applicable only to single excitations.* Although this method can be extended to a form that is applicable to double and more excitations, this is too time-consuming to be practical at present. Fortunately, *in large and low-symmetry systems such as those found naturally, significant excitations tend to be occupied by charge transfer ones, which are correctly represented by single excitations.* This contributes to the fact that this time-dependent response Kohn–Sham method is used in more than half of published excited state calculations (as of 2013). Note, however, that *using most functionals, the time-dependent Kohn–Sham method significantly underestimates charge transfer excitation energies.* This underesimation is solved by the long-range correction for exchange functionals (see Sect. 6.1).

4.7 Coupled Perturbed Kohn–Sham Method

Next, let us consider how to calculate chemical properties other than excitation energies. The key is that *most spectroscopic properties are response properties, which are proportional to the derivatives of energy in terms of various perturbations* (Jensen 2006),

Table 4.1 Major response properties assigned by n_F, n_B, n_I, and n_R in Eq. (4.44) (excerpt from (Jensen 2006))

n_F	n_B	n_I	n_R	Response property
0	0	0	0	Energy
1	0	0	0	Electric dipole moment
0	1	0	0	Magnetic dipole moment
0	0	1	0	Hyperfine coupling constant
0	0	0	1	Energy gradient
2	0	0	0	Electric polarizability
0	2	0	0	Magnetizability
0	0	2	0	Nuclear spin–spin coupling
0	0	0	2	Harmonic frequency
3	0	0	0	Hyperpolarizability
0	3	0	0	Hypermagnetizability
0	0	0	3	Anharmonic corrections
1	0	0	1	Infrared (IR) absorption intensity
1	1	0	0	Optical rotation circular dichloism (CD)
0	1	1	0	Nuclear magnetic shielding
2	0	0	1	Raman intensity
3	0	0	1	Hyper-Raman effect
2	1	0	0	Magnetic CD (Faraday effect)
1	0	0	2	IR intensity for overtone and combination bands
2	0	0	2	Raman intensity for overtone and combination bands
2	2	0	0	Cotton–Mouton effect

$$\text{Response property} \propto \frac{\partial^{n_F+n_B+n_I+n_R} E}{\partial \mathbf{F}_{\text{fld}}^{n_F} \partial \mathbf{B}_{\text{fld}}^{n_B} \partial \mathbf{I}_{\text{fld}}^{n_I} \partial \mathbf{R}^{n_R}}, \tag{4.44}$$

where $\mathbf{F}_{\text{fld}}$, $\mathbf{B}_{\text{fld}}$, $\mathbf{I}_{\text{fld}}$, and $\mathbf{R}$ are the perturbations of electric field, external magnetic field, internal magnetic field, and nuclear-coordinate vector, and n_X is the order of X perturbation. In Table 4.1, the orders of several major chemical properties are listed. As shown in the table, although many response properties are mixed derivatives depending on multiple perturbations, most properties can be described as energy derivatives up to the second order for each perturbation. Therefore, the energy derivatives are required only up to the second order to obtain these major response properties.

Energy derivatives are represented as perturbation terms. Assume the perturbed Hamiltonian operator of the Kohn–Sham method as

$$\hat{H} = \hat{H}_{\text{KS}} + \lambda V_1 + \lambda^2 V_2. \tag{4.45}$$

Based on the Rayleigh–Schrödinger perturbation theory (Schrödinger 1926), the first and second energy derivatives are written using the notation in Sect. 3.4 as

$$\left.\frac{\partial E_{\mathrm{KS}}}{\partial\lambda}\right|_{\lambda=0} = \langle \Psi_{\mathrm{KS}} | V_1 | \Psi_{\mathrm{KS}} \rangle + 2\left\langle \frac{\partial \Psi_{\mathrm{KS}}}{\partial\lambda} \middle| \hat{H}_{\mathrm{KS}} \middle| \Psi_{\mathrm{KS}} \right\rangle$$
$$= \langle \Psi_{\mathrm{KS}} | V_1 | \Psi_{\mathrm{KS}} \rangle , \tag{4.46}$$

and

$$\left.\frac{1}{2}\frac{\partial^2 E_{\mathrm{KS}}}{\partial\lambda^2}\right|_{\lambda=0} = \left\langle \frac{\partial \Psi_{\mathrm{KS}}}{\partial\lambda} \middle| V_1 \middle| \Psi_{\mathrm{KS}} \right\rangle + \langle \Psi_{\mathrm{KS}} | V_2 | \Psi_{\mathrm{KS}} \rangle , \tag{4.47}$$

respectively. The Hellmann–Feynman theorem (Feynman 1939) is used in Eq. (4.46). In these equations, only the first derivative of the Kohn–Sham wavefunction in terms of a perturbation, $\partial\Psi_{\mathrm{KS}}/\partial\lambda$, is not prepared in advance to calculate the energy derivatives up to the second order. It is also proven that higher than third energy derivatives are also represented with the wavefunction derivatives up to the first order. Therefore, only the first wavefunction derivatives are required to calculate the response properties. Based on perturbation theory, the first derivative of the Kohn–Sham wavefunction is represented as

$$\frac{\partial \Psi_{\mathrm{KS}}}{\partial\lambda} = \sum_{I}^{n_{\mathrm{exc.\ CSF}}} C_I \Psi_I , \tag{4.48}$$

where

$$C_I = \frac{\langle \Psi_I | V_1 | \Psi_{\mathrm{KS}} \rangle}{E_{\mathrm{KS}} - E_I} \tag{4.49}$$

However, it seems impossible to obtain the derivative exactly, because Eq. (4.48) contains the sum for all excitations created from a Kohn–Sham electron configuration. What makes it possible is the *coupled perturbed Kohn–Sham method.*

In the coupled perturbed Kohn–Sham method, the first wavefunction derivatives are given by calculating the first derivatives of the orbitals in terms of perturbations. The Kohn–Sham method is based on the Slater determinant. Therefore, since the Kohn–Sham wavefunction is represented with orbitals, the corresponding first wavefunction derivatives are also described by the first derivatives of the orbitals. For simplicity, let us consider the Kohn–Sham–Roothaan equation in Eq. (4.13), which is a matrix equation using basis functions based on the Roothaan method,

$$\mathbf{FC} = \mathbf{SC}\boldsymbol{\epsilon} , \tag{4.50}$$

and

$$\mathbf{C}^t\mathbf{SC} = \mathbf{1} . \tag{4.51}$$

The first derivative of Eq. (4.51) is given as

$$\mathbf{F}'\mathbf{C} + \mathbf{F}\mathbf{C}' = \mathbf{S}'\mathbf{C}\boldsymbol{\epsilon} + \mathbf{S}\mathbf{C}'\boldsymbol{\epsilon} + \mathbf{S}\mathbf{C}\boldsymbol{\epsilon}', \tag{4.52}$$

where

$$\mathbf{F}' = \mathbf{h}' + \left(\mathbf{P}'\mathbf{J} + \mathbf{P}\mathbf{J}'\right) + \mathbf{V}_{\mathrm{xc}}{}', \tag{4.53}$$

In these equations, the prime, "′", indicates the first derivative in terms of a perturbation and $\mathbf{P}$ is the density matrix. $\mathbf{J}$ is the Coulomb potential matrix containing

$$J_{pq} = \sum_{r,s}^{n_{\mathrm{basis}}} \langle pr|qs\rangle. \tag{4.54}$$

The first-order perturbation expansion of Eq. (4.52) is written as

$$\left[\mathbf{F}' - \mathbf{S}\boldsymbol{\epsilon}\right]\mathbf{C}' = \left[-\mathbf{F}' + \mathbf{S}'\boldsymbol{\epsilon} + \mathbf{S}\boldsymbol{\epsilon}'\right]\mathbf{C}, \tag{4.55}$$

under the orthonormalization condition,

$$\left(\mathbf{C}^t\mathbf{S}\mathbf{C}\right)' = \mathbf{0}. \tag{4.56}$$

This equation gives the first derivatives of the orbital cofficients, $\mathbf{C}'$, for calculating the first derivatives of the orbitals.

Considering the rotation (unitary transformation) of the orbitals makes it straightforward to solve the first derivatives of these orbitals. In the Kohn–Sham method, the orbitals and orbital energies are obtained by diagonalizing the Fock matrix. Therefore, the nondiagonal terms of the Fock matrix are zero for the oribtals, $\{\phi_i\}$,

$$\begin{aligned} F_{ia} &= \int d^3\mathbf{r}\phi_i^*(\mathbf{r})\hat{F}\phi_a(\mathbf{r}) \\ &= h_{ia} + \sum_j^{n_{\mathrm{orb}}} \langle ij|aj\rangle + (V_{\mathrm{xc}})_{ia} = 0, \end{aligned} \tag{4.57}$$

where

$$(V_{\mathrm{xc}})_{ia} = \int d^3\mathbf{r}\phi_i^*(\mathbf{r})V_{\mathrm{xc}}\phi_a(\mathbf{r}). \tag{4.58}$$

Since the Kohn–Sham method is based on the Slater determinant, orbital variation under perturbation is represented by the unitary transformation,

$$\phi_i' = \sum_j^{n_{\mathrm{basis}}} U_{ji}'\phi_j. \tag{4.59}$$

This is also represented as the variation of the orbital coefficient matrix, **C**, such as

$$\mathbf{C}' = \mathbf{U}'\mathbf{C}. \tag{4.60}$$

Substituting Eq. (4.59) into the first derivative of Eq. (4.57) leads to an equation similar to the time-dependent Kohn–Sham equation,

$$\mathbf{A}\mathbf{U}' = \mathbf{F}', \tag{4.61}$$

where matrix **A** contains

$$A_{ia,jb} = \delta_{ij}\delta_{ab}\left(\epsilon_a - \epsilon_i\right) + K_{ia,jb}. \tag{4.62}$$

$K_{ia,jb}$ corresponds to $K_{ia,jb}^{\sigma\tau}$ in Eq. (4.41), while neglecting spins. Note that this equation is based on the *Tamm–Dancoff approximation* (Hirata and Head-Gordon 1999), which is usually used in this method for simplicity. In Eq. (4.61), $\mathbf{F}'$ is the first derivative matrix of the Fock operator, which is usually derived as the first derivative of the one-electron parts, $\mathbf{h}'$ (McWeeny 1992). For the perturbation of a uniform electric field, $\mathbf{F}_{\text{fld}} = -F_{\text{fld}}\mathbf{r}$, this is given as the matrix containing (Lee and Colwell 1994)

$$F'_{ia} = h'_{ia} = \left(\frac{\partial \mathbf{h}}{\partial F_{\text{fld}}}\right)_{ia} = -\int d^3\mathbf{r}\phi_i^*(\mathbf{r})\mathbf{r}\phi_a(\mathbf{r}). \tag{4.63}$$

Using Eqs. (4.61) and (4.63), matrix $\mathbf{U}'$ is calculated to give the response properties in terms of the uniform electric field: dipole moments, polarizabilities, hyperpolarizabilities, and so forth. Equation (4.61) is called the *coupled perturbed Kohn–Sham equation*. Other response properties are calculated by solving Eq. (4.61) after setting the first derivative of the Fock operator, $\mathbf{F}'$, in terms of each perturbation. Note, however, that this method has problems in actual calculations similarly to the time-dependent response Kohn–Sham method. For example, *using most functionals, this method tends to overestimate the electric field response properties of long-chain polyenes*.

As another major method for calculating response properties, the *finite-field method* provides a useful example. This method directly estimates response properties by numerically calculating the infinitesimal energy variation resulting from adding a minite perturbation to a field. Although this direct method is more primitive than the coupled perturbed Kohn–Sham method, it is much used due to its equivalent results and easy-to-understand procedure. However, this method also overestimates the electric field response properties, similarly to the coupled perturbed Kohn–Sham method. Interestingly, *this overestimation is also solved by the long-range correction for exchange functionals* (see Sect. 6.1), analogously to the underestimation of charge transfer excitation energies mentioned in the previous section.

References

Casida, M.E.: In: Seminario, J.J. (ed.) Recent Developments and Applications of Modern Density Functional Theory. Elsevier, Amsterdam (1996)

Dirac, P.A.M.: Camb. Phil. Soc. **26**, 376–385 (1930)

Eschrig, H.: The Fundamentals of Density Functional Theory, 2nd edn. EAGLE, Leipzig (2003)

Fermi, E.: Z. Phys. **48**, 73–79 (1928)

Fermi, E., Amaldi E.: Accad. Ital. Rome **6**, 117–149 (1934)

Feynman, R.P.: Phys. Rev. **56**, 340–343 (1939)

Fock V.: Z. Phys. **61**, 126–148 (1930)

Gilbert, T.L.: Phys. Rev. B **12**, 2111–2120 (1975)

Gross, E.K.U., Burke, K.: Lect. Notes Phys. **706**, 1–17 (2006)

Gross, E.K.U., Ullrich, C.A., Gossmann, U.A.: In: Dreizler, R., Gross, E.K.U. (eds.), Density Functional Theory, NATO ASI Series B. Plenum, New York (1995)

Hirata, S., Head-Gordon, M.: Chem. Phys. Lett. **314**, 291–299 (1999)

Hohenberg, P., Kohn, W.: Phys. Rev. B **136**, 864–871 (1964)

Jensen F.: Introduction to Computational Chemistry. Wiley, Chichester (2006)

Kohn, W., Sham, L.J.: Phys. Rev. A **140**, 1133–1138 (1965)

Kutzelnigg, W.: J. Mol. Struct. Theochem **768**, 163–173 (2006)

Lee, A.M., Colwell, S.M.: J. Chem. Phys. **101**, 9704–9709 (1994)

Levy, M.: Proc. Natl. Acad. Sci. USA **76**, 6062–6065 (1979)

Lieb, E.H.: Int. J. Quantum Chem. **24**, 243–277 (1983)

Marques M.A.L., Castro, A., Rubio, A.: J. Chem. Phys. **115**, 3006–3014 (2001)

McWeeny, R.: Methods of Molecular Quantum Mechanics, 2nd edn. Academic Press, San Diego (1992)

Parr, R.G., Yang, W.: Density-Functional Theory of Atoms and Molecules. Oxford University Press, New York (1994)

Runge, E., Gross, E.K.U.: Phys. Rev. Lett. **52**, 997–1000 (1984)

Schipper, P.R.T., Gritsenko, O.V., Baerends, E.J.: Phys. Rev. A **57**, 1729–1742 (1998)

Schrödinger, E.: Ann. Phys. **80**, 437–490 (1926)

Seidl, A., Görling, A., Vogl, P., Majewski, J.A., Levy, M.: Phys. Rev. B **53**, 3764–3774 (1986)

Thomas, L.H.: Proc. Cam. Phyl. Soc. **23**, 542–548 (1927)

Ullrich, C.A.: Time-Dependent Density-Functional Theory. Oxford University Press, New York (2012)

van Leeuwen, R.: Lect. Notes Phys. **706**, 17–31 (2006)

von Weizsäcker, C.F.: Z. Phys. **96**, 431–458 (1935)

Yabana, K., Bertsch, G.F.: Phys. Rev. B **54**, 4484–4487 (1996)

Yang, Z., Burke, K.: Phys. Rev. A **88**, 042514(1–14) (2013)

Zhao, Q., Morrison, R.C., Parr, R.G.: Phys. Rev. A **50**, 2138–2142 (1994)

Chapter 5
Exchange-Correlation Functionals

5.1 Classification of Exchange-Correlation Functionals

In the previous chapter, the fundamentals of the Kohn–Sham method and its derivative theories have been explained without referring to the specific forms of exchange-correlation functionals used. The Kohn–Sham method is an established quantum theory based not only on electron density but also on a rigorous exchange-correlation functional. It is therefore difficult to assess the reliability of the Kohn–Sham method and its derivative theories without specifically considering the exchange-correlation functional used. In this chapter, let us examine the exchange-correlation functionals that have thus far been developed, with their features and problems.

As shown in Eq. (4.6), the exchange-correlation functional is the only part that is approximated in the Kohn–Sham equation. Therefore, the reliability of the Kohn–Sham method depends on the validity of this approximated functional. A variety of exchange-correlation functionals have thus far been developed on the basis of different physical models. In Fig. 5.1, exchange-correlation functionals are classified based on their characteristics.

- *Local density approximation (LDA) functionals*: Functionals of only electron density ρ.
- *Generalized gradient approximation (GGA) functionals*: Functionals correcting LDA functionals with the density gradient $\nabla\rho$.
- *Meta-GGA functionals*: Functionals correcting GGA functionals with the kinetic energy density τ.
- *Hybrid functionals*: Functionals mixing the Hartree–Fock exchange integral (E_x^{HF}) at a constant ratio.
- *Semiempirical functionals*: Functionals that are developed to reproduce accurate properties with many semiempirical parameters.
- *Progressive functionals*: Functionals transforming in accordance with combined functionals.

T. Tsuneda, *Density Functional Theory in Quantum Chemistry*,
DOI 10.1007/978-4-431-54825-6_5,

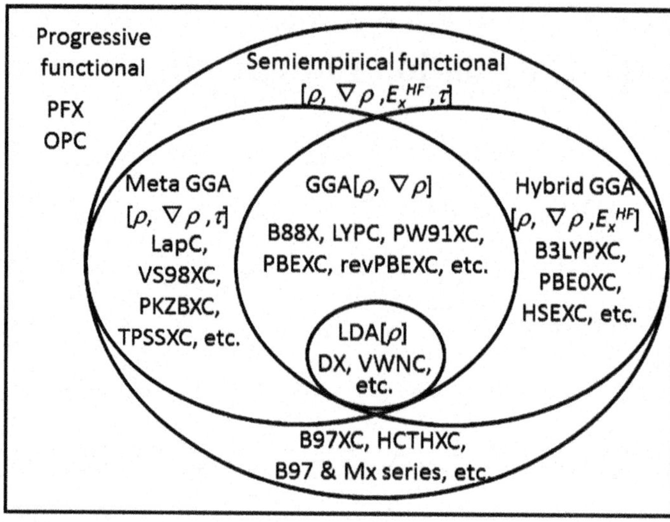

Fig. 5.1 Classification of major exchange-correlation functionals. The suffix "X" and "C" indicate exchange and correlation functionals, respectively

Although there are exchange-correlation functionals that do not fall into this classification, they are mostly functionals correcting the above functionals for some sort of physical factor. Such physical factors for functionals are introduced in Chap. 6. In this chapter, let us focus on functionals conforming to a standard paradigm in which LDA and GGA functionals are expanded by additional terms. These standard functionals are often arranged hierarchically, as in *Jacob's ladder* ("the ladder to Heaven" in *the Old Testament, Book of Genesis*, Fig. 5.2) (Perdew et al. 2006).

In past developments of exchange-correlation functionals, the following two criteria have been emphasized:

1. *Satisfying fundamental physical conditions.* Functionals have been physically assessed on the degree to which they satisfy fundamental physical conditions for each energy component (see Chap. 8).
2. *Accurately reproducing various reactions and properties for a wide range of molecules.* Functionals have been numerically assessed on the reproducibility of reaction diagrams and spectroscopic constants of molecules.

These two clear criteria have contributed to the active development of functionals. However, as a trade-off, brought about by overemphasizing these criteria for many years, we find an enormous number of functionals that attempt to enhance the physical validity with artificial additional terms or to improve the numerical accuracies with many semiempirical parameters. To solve this problem, we need the following new criteria for developing functionals:

3. *Simplicity, with a minimum number of parameters.* This clarifies the physical meaning of the functionals and makes it straightforward to explain the calculated results.

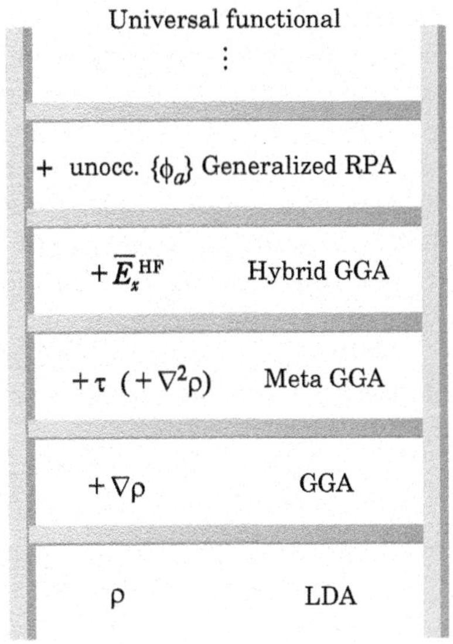

Fig. 5.2 Jacob's ladder in the development of functionals (Perdew et al. 2006)

4. *Containing no artificial terms added only to satisfy specific fundamental conditions or physical properties.* Improvements by superficial operations narrow the versatility and applicability of functionals by contrast.
5. *Taking into account physical corrections with no additional operation.* If functionals lack the versatility to take into account all physical corrections, like those given in Chap. 6 without modifying parameters, they would fade out in a short time.

In this chapter, exchange-correlation functionals are classified by their formulations, and the features and problems of these functionals are considered on the basis of the above criteria. In particular, the numbers of semiempirical parameters are mentioned. Since an enormous number of exchange-correlation functionals have been developed thus far, it is difficult to introduce all of them encyclopaedically. I therefore focus on the major functionals, which have taken into account the physical corrections given in Chap. 6 and are implemented in the predominantly used quantum chemistry calculation programs, e.g., Gaussian09 and GAMESS (current 2014).

5.2 LDA and GGA Exchange Functionals

The *LDA exchange functional has the exact form* in Eq. (4.3), which is the *Dirac LDA exchange functional.* Although this functional was developed to be used in the Thomas–Fermi method, it had been forgotten for many years. In 1951, Slater used this LDA exchange functional as an approximation for the exchange integral in the Hartree–Fock method (Slater 1951). Similarly to the usual exchange integral, this functional was first employed as an exchange potential of the Fock operator, and as a consequence it had significantly underestimated the exchange energies. To improve this underestimation, he suggested the *Xα method*, which multiplies a semiempirical parameter α by the exchange potential. This Xα method has been used in DFT in the field of solid state physics. On the other hand, the exact form of the Dirac LDA exchange functional is usually used in the Kohn–Sham method. Since the Dirac LDA exchange functional is also the exact local density approximation of exchange energy, it is usually used as the local density limit of exchange functionals (see Chap. 8).

GGA exchange functionals traditionally have a general form using a dimensionless coefficient K_σ, which is defined by

$$E_x = -\frac{1}{2}\sum_\sigma \int \rho_\sigma^{4/3} K_\sigma d^3\mathbf{r}. \tag{5.1}$$

This K_σ is usually expressed with the dimensionless parameter x_σ,

$$x_\sigma = \frac{|\nabla \rho_\sigma|}{\rho_\sigma^{4/3}}. \tag{5.2}$$

Although various forms of GGA exchange functionals have been developed, the differences in the best-known functionals are found only in the dependence of K_σ on x_σ. Figure 5.3 illustrates the dependence of K_σ on x_σ. The figure clearly shows that remarkable differences in GGA exchange functionals exist for large x_σ. That is, GGA exchange functionals are characterized by their differences in the region of low electron density and/or high density gradient. This remarkable difference is attributed to the lack of fundamental physical conditions for exchange energy in the low density/high density gradient region, in contrast to the exact local density and the generalized gradient approximation limits of exchange energy, which control the high density/low density gradient regions (see Chap. 8). Since this indicates that the exchange energy in the low density/high density gradient region is restricted by no fundamental physical condition, the exchange functional forms in these regions have been determined to improve the reproducibilities of properties. As mentioned above, there are many kinds of GGA exchange functionals. It is therefore difficult even to introduce only characteristic functionals. Figure 5.3 compares K_σ values for representative exchange functionals: the Becke 1988 (B88) functional (Becke 1988), the Perdew–Wang 1991 (PW91) functional (Perdew 1991; Perdew and Wang 1992), the Perdew–Burke–Ernzerhof (PBE) functional (Perdew et al. 1996), and the revised PBE (revPBE) functional (Zhang and Yang 1998).

The *B88 exchange functional* (Becke 1988) is the most popular GGA exchange functional in quantum chemistry calculations at present. This functional was derived for the form in the low density/high density gradient region to satisfy the *far-from-nucleus (long-range) asymptotic interaction condition* (see Chap. 8) such as

$$K_\sigma^{\mathrm{B88}} = K_\sigma^{\mathrm{LDA}} + \frac{2\zeta x_\sigma^2}{1 + 6\zeta x_\sigma \sinh^{-1} x_\sigma}, \tag{5.3}$$

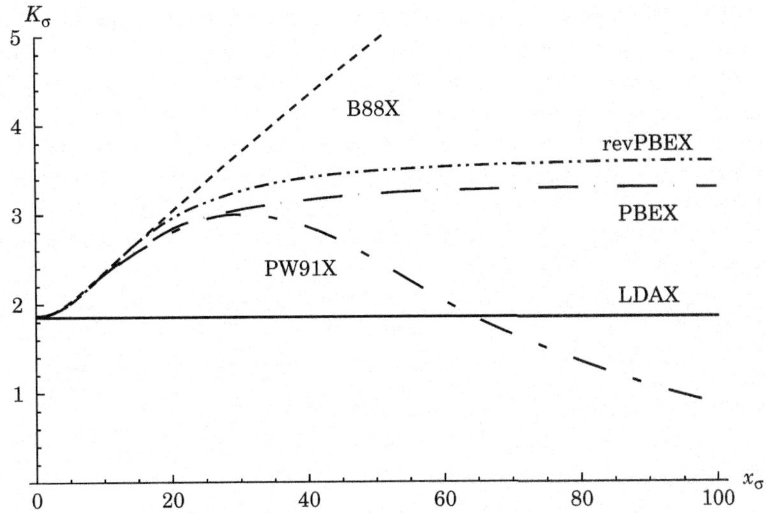

Fig. 5.3 The plot of K_σ, defined in Eq. (5.1), in terms of x_σ in Eq. (5.2) for major GGA exchange functionals

where

$$K_\sigma^{\mathrm{LDA}} = 3\left(\frac{3}{4\pi}\right)^{1/3}, \tag{5.4}$$

and the only parameter is $\zeta = 0.0042$. The feature of this functional is in the high reproducibility of the exchange energies. In particular, this functional yielded much more accurate exchange energies of atoms and molecules than those of other functionals that had been developed previously. The high accuracy of this functional led to the explosive expansion in the use of DFT in quantum chemistry calculations. However, since this functional takes no fundamental physical condition other than the long-range (far-nucleus) asymptotic interaction condition into consideration, the physical validity of this functional is not so high (see Sect. 8).

The *PW91 exchange functional* (Perdew 1991; Perdew and Wang 1992) is a reformulation of the B88 exchange functional designed to satisfy as many fundamental physical conditions as possible:

$$K_\sigma^{\mathrm{PW91}} = K_\sigma^{\mathrm{LDA}}\left[\frac{1+6\zeta x_\sigma \sinh^{-1} x_\sigma}{1+6\zeta x_\sigma \sinh^{-1} x_\sigma + 0.004x_\sigma^4/(48\pi^2)^{4/3}}\right.$$
$$\left.+\frac{(0.2743-0.1508\exp[-100x_\sigma^2/(48\pi^2)^{2/3}])x_\sigma^2/(48\pi^2)^{2/3}}{1+6\zeta x_\sigma \sinh^{-1} x_\sigma + 0.004x_\sigma^4/(48\pi^2)^{4/3}}\right]. \tag{5.5}$$

This functional contains an additional term and four parameters to satisfy fundamental physical conditions and one parameter ζ originally from the B88 functional. Notice that the parameters fitted to satisfy fundamental physical conditions are called *fundamental constants* and are often regarded as being intrinsically different from semiempirical parameters, despite the fact they are not included in the original universal physical constants such as the Planck constant. The feature of this functional is to obey more fundamental physical conditions than the B88 functional does. However, in exchange, it has a complicated form. As shown in Fig. 5.3, this functional also produces an exchange energy close to zero for large x_σ in Eq. (5.2). This behavior often causes problems in property calculations.

The *PBE exchange functional* (Perdew et al. 1996) simplifies the PW91 exchange functional by restricting applied fundamental physical conditions to significant ones and by using only two fundamental constants. The K_σ of this functional is

$$K_\sigma^{\rm PBE} = K_\sigma^{\rm LDA}\left[1+\kappa-\frac{\kappa}{1+\mu x_\sigma^2/(48\pi^2)^{2/3}\kappa}\right], \tag{5.6}$$

where the fundamental constants are $\mu = 0.21951$ and $\kappa = 0.804$. As shown in Fig. 5.3, since this K_σ approaches a constant, this functional hardly contributes to the exchange energy for large x_σ (i.e., in the low density/high density gradient region). Although this behavior may not be correct from a physical viewpoint, it seems efficient for weakly bonded and high-spin systems. The *revPBE exchange functional* (Zhang and Yang 1998) only modifies the parameters of this PBE exchange functional ($\mu = 0.967$ and $\kappa = 0.235$), and therefore it has the same characteristics as those of this PBE functional.

As described thus far, the mainstream GGA exchange functionals are the B88 exchange functional and its modifications for satisfying fundamental physical conditions at present. Now let us examine an exchange functional that was developed from a different standpoint, the *parameter-free* (PF) *exchange functional* (Tsuneda and Hirao 2000). This functional is directly derived from a density matrix expansion at the Fermi momentum (Negele and Vautherin 1972),

$$\begin{aligned} P_\sigma(\mathbf{r}_1,\mathbf{r}_2) &= \frac{3j_1(k_{{\rm F}\sigma}r_{12})}{k_{{\rm F}\sigma}r_{12}}\rho_\sigma(\mathbf{r}) + \frac{35j_3(k_{{\rm F}\sigma}r_{12})}{2k_{{\rm F}\sigma}^3 r_{12}}\left(\frac{\nabla^2\rho_\sigma(\mathbf{r})}{4} - 2\tau_\sigma(\mathbf{r}) + \frac{3}{5}k_{{\rm F}\sigma}^2\rho_\sigma(\mathbf{r})\right) \\ &\quad + \cdots, \end{aligned} \tag{5.7}$$

where $\mathbf{r} = (\mathbf{r}_1 + \mathbf{r}_2)/2$, j_n is the spherical Bessel function, and

$$k_{{\rm F}\sigma} = (6\pi^2\rho_\sigma)^{1/3} \tag{5.8}$$

is the Fermi momentum that indicates the largest momentum of electronic motions in momentum space. In the derivation of this functional, the Fermi momentum is calculated using the kinetic energy density τ_σ such as

$$k_{F\sigma} = \sqrt{\frac{10\tau_\sigma}{3\rho_\sigma}}. \tag{5.9}$$

As a result, this functional is in a form containing the kinetic energy density,

$$K_\sigma^{PF} = \frac{27\pi}{10\tau_\sigma}\rho_\sigma^{5/3}\left(1 + \frac{7x_\sigma^2\rho_\sigma^{5/3}}{216\tau_\sigma}\right). \tag{5.10}$$

Since this functional has neither semiempirical parameters nor fundamental constants, the kinetic energy density is the only adjustable part. If the exact independent electron approximation kinetic energy density is used for this kinetic energy density part, this functional usually gives a divergent energy. However, if the Thomas–Fermi–Weizsäcker (TFW) GGA kinetic energy functional, which is the exact density gradient expansion of kinetic energy, is applied, this functional reproduces values of the atomic exchange energies that are on average 97.8% (mean absolute error 0.154 atomic units) of the exact values for the elements H through Ar, in spite of its parameterless form (Tsuneda and Hirao 2000). Even more surprisingly, it was proved that this functional satisfies most fundamental physical conditions if the fundamental physical conditions for kinetic energy are applied to the kinetic energy density part (see Chap. 8). Notice that although this functional is explained in the section on GGA exchange functionals, it is actually a *progressive functional*, transforming in accordance with a combined kinetic energy functional, for which any type of functional is available.

5.3 LDA and GGA Correlation Functionals

In contrast to exchange functionals, *there is no exact LDA correlation functional.* The analytical derivation of the exact LDA correlation functional has been examined for many years. Consequently, the high density limit (Gell-Mann and Bruecker 1957), low density limit (Carr 1961), and random phase approximation (RPA) expression (Hedin and Lundqvist 1971) have been suggested for the LDA correlation functional. However, the high and low density limits provide no information on the LDA correlation functional in the realistic mid-range density regions, and the RPA expression does not give these limits. As alternates, various approximate LDA correlation functionals have been suggested: The most-used LDA correlation functionals are the Vosko–Wilk–Nusair (VWN) functional, (Vosko et al. 1980) mainly used in quantum chemistry calculations and the Perdew–Wang functional (Perdew and Wang 1992), mainly used in solid state calculations.

The *VWN LDA correlation functional* (Vosko et al. 1980) was developed to make the RPA expression converge to the high and low density limits. This functional was inductively derived on the basis of the Padè interpolation by fitting parameters to the exact correlation energy of a uniform density gas given by the quantum Monte

Carlo method, such as

$$
\begin{aligned}
&E_{\mathrm{c}}^{\mathrm{VWN}} \\
&= \int d^3\mathbf{r} \left\{\epsilon_0^{\mathrm{VWN}}(x_{01}, a_1, b_1, c_1)\right. \\
&+\epsilon_0^{\mathrm{VWN}}(x_{02}, a_2, b_2, c_2)\left(\frac{9(f_1[\zeta]-2)}{2}\right)(1-\zeta^4) \\
&\left.+\left[\epsilon_0^{\mathrm{VWN}}(x_{03}, a_3, b_3, c_3)-\epsilon_0^{\mathrm{VWN}}(x_{01}, a_1, b_1, c_1)\right]\frac{f_1(\zeta)-2}{2^{1/3}-1}\zeta^4\right\},
\end{aligned}
\tag{5.11}
$$

where $\epsilon_0^{\mathrm{VWN}}$ is expressed as

$$
\begin{aligned}
&\epsilon_0^{\mathrm{VWN}}(x_0, a, b, c) \\
&= \frac{a}{2}\left\{\ln\left(\frac{r_{\mathrm{s}}}{r_{\mathrm{s}}+b\sqrt{r_{\mathrm{s}}}+c}\right)+\frac{2b}{\sqrt{4c-b^2}}\tan^{-1}\left(\frac{\sqrt{4c-b^2}}{2\sqrt{r_{\mathrm{s}}}+b}\right)\right. \\
&-\frac{bx_0}{x_0^2+bx_0+c}\left[\ln\left(\frac{(\sqrt{r_{\mathrm{s}}}-x_0)^2}{r_{\mathrm{s}}+b\sqrt{r_{\mathrm{s}}}+c}\right)\right. \\
&\left.\left.+\frac{2(b+2x_0)}{\sqrt{4c-b^2}}\tan^{-1}\left(\frac{\sqrt{4c-b^2}}{2\sqrt{r_{\mathrm{s}}}+b}\right)\right]\right\}.
\end{aligned}
\tag{5.12}
$$

and $f_1[\zeta]$ and ζ are given by

$$
f_1[\zeta] = \frac{(1+\zeta)^{4/3}+(1-\zeta)^{4/3}}{2} \tag{5.13}
$$

and

$$
\zeta = \frac{\rho_\alpha - \rho_\beta}{\rho_\alpha + \rho_\beta}. \tag{5.14}
$$

In Eq. (5.12), r_{s} is Wigner–Seitz radius defined by

$$
\frac{4}{3}\pi r_{\mathrm{s}}^3 = \frac{1}{\rho}. \tag{5.15}
$$

The VWN LDA correlation functional has 12 parameters in total: x_{0i}, a_i, b_i, c_i for $i = 1-3$. Notice that there are five combinations for these parameters (ver. 5 is usually employed. For the parameter values, see Vosko et al. 1980).

The *PW LDA correlation functional* (Perdew and Wang 1992) reforms the VWN LDA correlation functional to simplify it, to reduce the number of parameters, and to satisfy more fundamental physical conditions. The functional form is

$$E_{\mathrm{c}}^{\text{PW-LDA}}[\rho] = -2a \int d^3\mathbf{r}\rho\,(1-\alpha r_{\mathrm{s}}) \ln\left[1+\frac{1}{2a\left(\beta_1 r_{\mathrm{s}}^{1/2}+\beta_2 r_{\mathrm{s}}+\beta_3 r_{\mathrm{s}}^{3/2}+\beta_4 r_{\mathrm{s}}^2\right)}\right], \tag{5.16}$$

where the parameters are 6 in total: $a = 0.031097$, $\alpha = 0.21370$, $\beta_1 = 7.5957$, $\beta_2 = 3.5876$, $\beta_3 = 1.6382$, and $\beta_4 = 0.49294$.

These LDA correlation functionals have been used in various property calculations, especially in solid state calculations. However, we should recall that these *LDA correlation functionals are not exact functionals but are inductively derived approximative functionals*. Actually, even though the exact correlation energy of a uniform electron gas in a quantum Monte Carlo calculation has $O(\rho)$ and $O(\rho^{4/3})$ density dependences at the high and low density limits, respectively (Ceperley and Alder 1980), these LDA correlation functionals have been proven to violate at least one of these limits (Tsuneda et al. 2001).

GGA correlation functionals are mainly classified into the *density gradient approximation-types* and the *Colle–Salvetti (CS)-type functionals*.

The *density gradient approximation-type correlation functionals have been developed in a similar manner to the GGA exchange functionals*. This type of correlation functional is generally derived by determining a basic form to satisfy some fundamental physical conditions of the correlation energy (see Chap. 8) and then fitting parameters to reproduce other conditions. In these types of functionals, the PW91 and PBE correlation functionals are most frequently used. These density gradient approximation-type correlation functionals have the feature of satisfying many fundamental physical conditions, but with the problem of complicated forms, with many fitted parameters (fundamental constants).

The *PW91 correlation functional* (Perdew and Wang 1992) augments the PW LDA correlation functional with a GGA term *using parameters fitted to satisfy as many fundamental physical conditions of the correlation energy as possible.* The resulting functional has a very complicated form,

$$E_{\mathrm{c}}^{\text{PW91}}[\rho,s,t] = E_{\mathrm{c}}^{\text{PW-LDA}}[\rho] + \int d^3\mathbf{r}\rho H[\rho,s,t] \tag{5.17}$$

$$\begin{aligned} H[\rho,s,t] = {} & \frac{\beta^2}{2\alpha}\ln\left[1+\frac{2\alpha}{\beta}\frac{t^2+At^4}{1+At^2+A^2t^4}\right] \\ & + C_{c_0}\left[C_1+\frac{C_2+C_3 r_{\mathrm{s}}+C_4 r_{\mathrm{s}}^2}{1+C_5 r_{\mathrm{s}}+C_6 r_{\mathrm{s}}^2+C_7 r_{\mathrm{s}}^3}-C_{c_1}\right] \\ & \times t^2\exp(-100s^2), \end{aligned} \tag{5.18}$$

where the coefficient A is

$$A = \frac{2\alpha}{\beta}\left[\exp\left(\frac{-2\alpha \bar{E}_{\mathrm{c}}^{\mathrm{PW\text{-}LDA}}[\rho]}{\beta^2\rho}\right) - 1\right]^{-1}, \tag{5.19}$$

and the dimensionless parameters s and t are

$$s = \frac{|\nabla\rho|}{2k_{\mathrm{F}}\rho} \text{ and } t = \frac{|\nabla\rho|}{2k_{\mathrm{s}}\rho}, \tag{5.20}$$

in which $k_{\mathrm{F}} = (3\pi^2\rho)^{1/3}$ and $k_{\mathrm{s}} = (4k_{\mathrm{F}}/\pi)^{1/2}$. $\bar{E}_{\mathrm{c}}^{\mathrm{PW\text{-}LDA}}$ is the integral kernel of the PW LDA correlation functional. In the GGA term of this functional, there are 11 fundamental constants in total, α, β, C_{c_0}, C_{c_1} and C_i for $i = 1 - 7$, which are determined to satisfy fundamental physical conditions (for parameter values, see Perdew et al. 1996). Although this functional satisfies a certain number of fundamental physical conditions, several problems have been indicated. One of the problems comes from the large number of parameters (fundamental constants), Perdew, a developer of this functional, suggests that this can cause problems in property calculations (Perdew et al. 1996). As a more severe essential problem, the fundamental physical condition of correlation energy for the low density/high density gradient limit (Ma and Brueckner 1968) (see Chap. 8), which is used in the derivation of this functional, is found to have a density dependence of the exchange energy by dimensional analysis. This is because this condition was derived based on the LDA exchange functional and therefore supplements the deficient exchange energy in this correlation functional. As a result, the *PW91 correlation functional contains the effect of exchange interactions.* It is expected as a counterargument that this effect can be interpreted as a nondynamical electron correlation. However, since the nondynamical electron correlation is usually included in exchange functionals (see Sect. 4.5), this interpretation may cause a new, awkward problem, i.e., doubly-counted electron correlations.

The *PBE correlation functional* (Perdew et al. 1996) was developed to solve various problems in the PW91 correlation functional by drastically reducing the parameters, from 11 to 2, and by focusing on three significant fundamental physical conditions to be satisfied. The functional form is expressed as

$$E_{\mathrm{c}}^{\mathrm{PBE}}[\rho,\zeta,t] = E_{\mathrm{c}}^{\mathrm{PW\text{-}LDA}}[\rho] + \int d^3\mathbf{r}\rho H[\rho,\zeta,t] \tag{5.21}$$

$$H[\rho,\zeta,t] = \gamma\phi^3 \ln\left[1 + \frac{\beta}{\gamma}t'^2\left(\frac{1 + At'^2}{1 + At'^2 + A^2t'^4}\right)\right], \tag{5.22}$$

$$A = \frac{\beta}{\gamma}\left[\exp\left(-\frac{\bar{E}_{\mathrm{c}}^{\mathrm{PW\text{-}LDA}}[\rho]}{\gamma\phi^2\rho}\right) - 1\right]^{-1} \tag{5.23}$$

$$\phi = \frac{1}{2}\left[(1+\zeta)^{2/3} + (1-\zeta)^{2/3}\right], \tag{5.24}$$

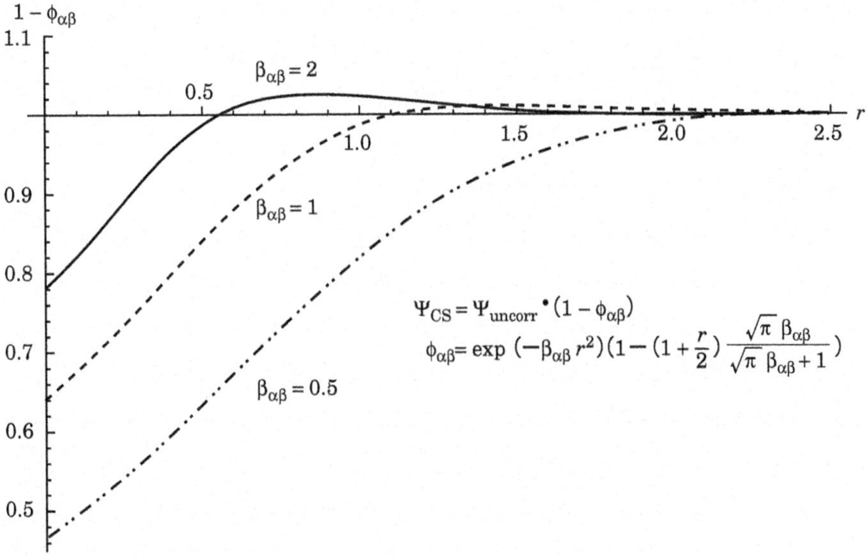

Fig. 5.4 Form of the correlation function for the Colle–Salvetti-type correlation functionals. Ψ_{uncorr} is a wavefunction containing no electron correlation effect

where ζ is given in Eq. (5.14). This functional is clearly superior to the PW91 correlation functional, because it is much simpler, with only two fundamental constants ($\gamma = (1 - \ln 2)/\pi^2 = 0.031091$ and $\beta = 0.066725$) in addition to those in the PW LDA functional. However, this functional also contains exchange effects due to the use of the low density/high density gradient limit condition (Ma and Brueckner 1968) mentioned above in the derivation. Since this limit condition is also inconsistent with the uniform coordinate scaling condition (see Chap. 8), *this functional has an unnatural form to overcome this inconsistency*.

The *CS-type correlation functional* (Colle and Salvetti 1975) is derived from a correlation wavefunction that multiplies an uncorrelated wavefunction with a correlation function, giving a correlation hole satisfying the *correlation cusp condition* for short-range electron–electron interactions, as mentioned in Sect. 3.2. Figure 5.4 illustrates the general form of the correlation function. In the CS-type functionals, the Lee–Yang–Parr (LYP) functional (Lee et al. 1988) and the one-parameter progressive (OP) functional (Tsuneda et al. 1999) are included. For these functionals, the features are their simple physical models, multiplying only correlation functions and their resulting highly accurate correlation energies for molecules. On the other hand, the CS-type functionals, except for the OP functional hardly satisfy fundamental physical conditions, in contrast to the density gradient approximation-type ones, for which these conditions are used in the derivations.

The original *CS correlation functional* was proposed as *an electron correlation correction for the Hartree–Fock method* in 1975. Assuming that the volume of the region where electrons are excluded (excluded volume) is proportional to Wigner's excluded volume (Wigner 1934; Wigner and Seitz 1933), the following equation was derived:

$$E_c^{CS}[\rho, \nabla_{\mathbf{r}}^2 P_{2HF}] = -\int d^3\mathbf{r} \frac{a}{1+d\beta} \rho \left[1 + bW \exp(-c/\beta)\right], \tag{5.25}$$

where

$$W = \rho^{-8/3} \left[\nabla_{\mathbf{r}}^2 P_{2HF}\left(\mathbf{r} - \frac{\mathbf{s}}{2}, \mathbf{r} + \frac{\mathbf{s}}{2}\right)\right]_{\mathbf{s}=0}, \tag{5.26}$$

and

$$\beta = q\rho^{1/3}. \tag{5.27}$$

In this functional, five semiempirical parameters, $a = 0.04918$, $b = 0.06598$, $c = 0.58$, $d = 0.8$, and $q = 2.29$, are used with no fundamental constant. This functional accurately reproduces the correlation energies of atoms and small molecules. However, the parameters, which are fitted to the Hartree–Fock wavefunction, should be modified for combining with an exchange functional in the Kohn–Sham method. The time-consuming second derivative term of the diagonal term of the second-order density matrix in Eq. (5.26) may also be a problem from the practical point of view.

The *LYP correlation functional* (Lee et al. 1988) made the CS functional suitable for practical use. In this functional, *the second derivative term* in Eq. (5.26) *is approximated by use only of a density gradient to obtain a GGA functional* containing only the density and density gradient. This approximation made it facile to implement this functional in computational programs and reduced the computational time. This functional is in the following form:

$$\begin{aligned} &E_c^{LYP}[\rho, \nabla\rho, \nabla^2\rho] \\ &= -\int d^3\mathbf{r} \frac{a}{1+d\rho^{-1/3}} \\ &\left\{\rho + b\rho^{-2/3}\left[C_F\rho^{5/3} - 2t_W + \frac{1}{9}\left(t_W + \frac{1}{2}\nabla^2\rho\right)\right]\exp(-c\rho^{-1/3})\right\}, \end{aligned} \tag{5.28}$$

where

$$t_W = \frac{1}{8}\left(\frac{|\nabla\rho|^2}{\rho} - \nabla^2\rho\right) \tag{5.29}$$

and β and C_F are given in Eqs. (5.27) and (4.2), respectively. In this functional, five semiempirical parameters, $a = 0.04918$, $b = 0.7628$, $c = 0.58$, $d = 0.8$, and $q =$

2.29 are contained, with no fundamental constant. Since this LYP correlation functional gives very accurate correlation energies in molecular property calculations, it is the most used correlation functional in quantum chemistry calculations (current 2014). Notice that this functional contains a term including a density Laplacian term $\nabla^2\rho$ and therefore seems not to be a GGA correlation functional. However, this density Laplacian term is easily transformed to a density gradient term via a partial integral. This functional has another severe problem that has been neglected: *the sign of the second derivative term is opposite to that of the CS functional* (Tsuneda et al. 1999). Although this opposite sign may be incorporated to obtain accurate correlation energies, it clearly undermines the physical validity of this functional. Actually, *this functional hardly satisfies fundamental physical conditions*, as shown in Chap. 8. This may be one of the main reasons for the fact that the LYP functional is hardly used in solid state calculations.

The *OP correlation functional* (Tsuneda et al. 1999) *determines the exclusion volume of the correlation hole, by assuming that it is proportional to the exclusion volume of the exchange hole coming from the combined exchange functional, in order to enhance the physical validity of the CS-type correlation functional.* In this functional, only electron correlations of opposite-spin pairs are explicitly incorporated, while electron correlations of parallel-spin pairs are secondary effects from the combined exchange functional. It has only one semiempirical parameter, which is the minimum required for adapting to the exchange functional. As a result, a simple correlation functional was derived:

$$E_c^{\mathrm{OP}} = -\int d^3\mathbf{r}\rho_\alpha\rho_\beta \frac{1.5214\beta_{\alpha\beta} + 0.5764}{\beta_{\alpha\beta}^4 + 1.1284\beta_{\alpha\beta}^3 + 0.3183\beta_{\alpha\beta}^2}, \tag{5.30}$$

where

$$\beta_{\alpha\beta} = q_{\alpha\beta}\left(\rho_\alpha^{-1/3}K_\alpha^{-1} + \rho_\beta^{-1/3}K_\beta^{-1}\right)^{-1}, \tag{5.31}$$

and K_σ is the exchange functional term defined in Eq. (5.1). The semiempirical parameter is only $q_{\alpha\beta}$, and no fundamental constant is included. The parameter $q_{\alpha\beta}$ is determined for each exchange functional combined: e.g., $q_{\alpha\beta} = 2.367$ if this functional is combined with the B88 and PBE exchange functionals. Since the OP functional contains no density gradient term, except for that in the exchange functional term K_σ, it is an LDA correlation functional for combining with the LDA exchange functional and a GGA correlation functional for combining with a GGA exchange functional. That is, *the OP functional is a progressive functional transforming in accordance with a combined exchange functional.* In spite of the OP functional being simple, with only one parameter, it gives correlation energies similar to those of the LYP correlation functional. Furthermore, *the OP functional satisfies the most fundamental physical conditions in all correlation functionals*, although it uses no such condition in the derivation, differently from density gradient approximation-type functionals (see Chap. 8).

In the CS-type correlation functionals, there are the *Lap-series meta-GGA correlation functionals* (see Sect. 5.4), among others.

5.4 Meta-GGA Functionals

The *meta-GGA functional improves the approximation of the GGA functional by using the kinetic energy density* τ. In the density matrix expansion around the Fermi momentum given in Eq. (5.7), the kinetic energy density, τ, and the density Laplacian, $\nabla^2\rho$, are included in the term following the density gradient term (Negele and Vautherin 1972). Since the exchange energy is represented by the density matrix, the kinetic energy density appears in the next correction. Note that the density Laplacian term follows the Thomas–Fermi (TF) LDA kinetic energy functional and the Weizsäcker correction term (see Sect. 4.1) in the kinetic energy density,

$$\tau = \sum_{\sigma} \tau_{\sigma} = \frac{1}{2}\sum_{\sigma}\sum_{i} |\nabla\phi_{i\sigma}|^2 \tag{5.32}$$

$$= \sum_{\sigma}\left[C_{\mathrm{F}}\rho_{\sigma}^{5/3} + \frac{1}{72}\frac{|\nabla\rho_{\sigma}|^2}{\rho_{\sigma}} + \frac{1}{6}\nabla^2\rho_{\sigma} + O(\nabla^4)\right], \tag{5.33}$$

where C_{F} is the coefficient of the spin-dependent TF LDA kinetic energy functional (Eq. (4.2)),

$$C_{\mathrm{F}} = \frac{3}{10}(6\pi^2)^{2/3}. \tag{5.34}$$

Therefore, the effect of the density Laplacian is included implicitly in the kinetic energy density. It is natural that the next step in density gradient correction is the kinetic energy density correction on Jacob's ladder (see Sect. 5.1). Major meta-GGA functionals include the van Voorhis–Scuseria 1998 (VS98) meta-GGA exchange-correlation (van Voorhis and Scuseria 1998), the Perdew–Kurt–Zupan–Blaha (PKZB) meta-GGA exchange-correlation (Perdew et al. 1999), and the Tao–Perdew–Staroverov–Scuseria (TPSS) meta-GGA exchange-correlation (Tao et al. 2003) functionals.

As far as I know, the first meta-GGA functional was the *Lap-series meta-correlation functional* (Proynov et al. 1995). This functional is also a CS-type correlation functional (see Sect. 5.3). In this functional, the size of the exclusion volume is represented by a momentum, which is written by the kinetic energy density to derive a CS-type correlation functional that depends on the kinetic energy density,

$$E_{c\alpha\beta}^{\mathrm{Lap1}} = \int d^3\mathbf{r}\rho_{\alpha}\rho_{\beta}Q_{\alpha\beta}, \tag{5.35}$$

and

$$E_{c\sigma\sigma}^{\text{Lap1}} = \int d^3\mathbf{r}\rho\left(1 - \frac{1}{n_\sigma}\right)\frac{\rho_\sigma^2}{\rho}C_{\text{p}}Q_{\sigma\sigma}, \tag{5.36}$$

where

$$Q_{\sigma\sigma'} = -\frac{b_1}{1 + b_2 k_{\sigma\sigma'}} + \frac{b_3}{k_{\sigma\sigma'}}\ln\left(\frac{b_4 + k_{\sigma\sigma'}}{k_{\sigma\sigma'}}\right) + \frac{b_5}{k_{\sigma\sigma'}} - \frac{b_6}{k_{\sigma\sigma'}^2}, \tag{5.37}$$

$$k_{\sigma\sigma'} = \frac{2k_\sigma k_{\sigma'}}{k_\sigma + k_{\sigma'}} \tag{5.38}$$

and

$$k_\sigma = \alpha_{\text{e}}\sqrt{\frac{2\tau_\sigma}{3\rho_\sigma}}. \tag{5.39}$$

In these equations, n_σ is the number of σ-electrons and eight semiempirical parameters are used: b_i $(i = 1 - 6)$, α_{e}, and C_{p} (for the parameter values, see Proynov et al. 1995). This is called the *Lap1 correlation functional.* Later, the *Lap3 correlation functional* (Proynov et al. 1997) was developed to enhance the Lap1 functional using the second-order density matrix. These functionals are of great significance in being the first to incorporate the kinetic energy density. However, they hardly satisfy fundamental conditions, similarly to most CS-type correlation functionals and contain many semiempirical parameters.

The *VS98 meta-GGA exchange-correlation functional* (van Voorhis and Scuseria 1998) is the first exchange functional incorporating the kinetic energy density. Similarly to the PF exchange functional (see Sect. 5.2), this functional is derived from the analytical expansion of the density matrix in Eq. (5.7) (Negele and Vautherin 1972),

$$E_{\text{x}}^{\text{VS98}} = \sum_\sigma \int d^3\mathbf{r}\rho_\sigma^{4/3} h_\sigma(x_\sigma, z_\sigma), \tag{5.40}$$

$$E_{c\sigma\sigma'}^{\text{VS98}} = \int d^3\mathbf{r}\hat{E}_{c\sigma\sigma'}^{\text{LDA}} h_\sigma\left(\sqrt{x_\alpha^2 + x_\beta^2}, z_\alpha + z_\beta\right), \tag{5.41}$$

and

$$E_{c\sigma\sigma}^{\text{VS98}} = \int d^3\mathbf{r}\hat{E}_{c\sigma\sigma}^{\text{LDA}} h_\sigma(x_\sigma, z_\sigma), \tag{5.42}$$

where

$$h_\sigma(x_\sigma, z_\sigma) = \frac{a}{\gamma_\sigma(x_\sigma, z_\sigma)} + \frac{bx_\sigma^2 + cz_\sigma}{\gamma_\sigma^2(x_\sigma, z_\sigma)} + \frac{dx_\sigma^4 + ex_\sigma^2 z_\sigma + fz_\sigma^2}{\gamma_\sigma^3(x_\sigma, z_\sigma)}, \tag{5.43}$$

and

$$\gamma_\sigma(x_\sigma, z_\sigma) = 1 + \alpha(x_\sigma^2 + z_\sigma). \tag{5.44}$$

$\hat{E}_{c\sigma_1\sigma_2}^{\mathrm{LDA}}$ is the integral kernel of the $\sigma_1\sigma_2$-pair LDA correlation functional. Both x_σ and z_σ are dimensionless parameters: x_σ is given in (5.2) and z_σ is defined using the kinetic energy density as

$$z_\sigma = \frac{\tau_\sigma - \tau_\sigma^{\mathrm{TF}}}{\rho_\sigma^{5/3}} = \frac{\tau_\sigma}{\rho_\sigma^{5/3}} - C_{\mathrm{F}}. \tag{5.45}$$

This functional contains 7 semiempirical parameters (a through f and α) for each exchange, parallel-spin or opposite-spin pair correlation functional, and consequently has 21 semiempirical parameters in total (for the parameter values, see van Voorhis and Scuseria 1998). Therefore, this functional is also classified as a semiempirical functional (see Sect. 5.6) and is actually taken as the first meta-GGA semiempirical functional using z_σ. This functional is also characteristic in associating the correlation functional with the exchange one through the same h_σ.

The term, "meta-GGA functional," first appeared in the *PKZB meta-GGA exchange-correlation functional* (Perdew et al. 1999). The PKZB exchange functional intends to enhance the PBE-GGA functional using the kinetic energy density on the basis of the fundamental conditions extended to the density Laplacian,

$$E_{\mathrm{x}}^{\mathrm{PKZB}} = \sum_\sigma \int d^3\mathbf{r} \bar{E}_{\mathrm{x}}^{\mathrm{LDA}}[2\rho_\sigma] \left\{ 1 + \kappa - \frac{\kappa}{1 + x/\kappa} \right\}, \tag{5.46}$$

where

$$x = \frac{10}{81}p + \frac{146}{2025}\tilde{q}^2 - \frac{73}{405}\tilde{q}p + \left[D + \frac{1}{\kappa}\left(\frac{10}{81}\right)^2 \right] p^2, \tag{5.47}$$

$$\tilde{q} = \frac{3\tau_\sigma}{2(3\pi^2)^{2/3}\rho_\sigma^{5/3}} - \frac{9}{20} - \frac{p}{12}, \tag{5.48}$$

and

$$p = \frac{|\nabla\rho_\sigma|^2}{4(6\pi^2)^{2/3}\rho_\sigma^{8/3}}. \tag{5.49}$$

In this functional, two semiempirical parameters, $D = 0.113$ and $\kappa = 0.804$, are included. The PKZB correlation functional removes the self-interaction error (see Sect. 6.2) of the PBE-GGA correlation functional to reproduce the one-electron self-correlation energy in the case that electrons interact only with themselves (i.e., $\tau_\sigma = \tau_\sigma^{\mathrm{W}}$),

$$E_{\mathrm{c}}^{\mathrm{PKZB}} = \int d^3\mathbf{r} \left\{ \bar{E}_{\mathrm{c}}^{\mathrm{PBE}}(\rho_\alpha, \rho_\beta, \nabla\rho_\alpha, \nabla\rho_\beta) \left[1 + C \left(\frac{\sum_\sigma \tau_\sigma^{\mathrm{W}}}{\sum_\sigma \tau_\sigma} \right)^2 \right] \right.$$
$$\left. -(1+C) \sum_\sigma \left(\frac{\tau_\sigma^{\mathrm{W}}}{\tau_\sigma} \right)^2 \bar{E}_{\mathrm{c}}^{\mathrm{PBE}}(\rho_\sigma, 0, \nabla\rho_\sigma, 0) \right\} , \tag{5.50}$$

where one semiempirical parameter, $C = 0.53$, is included. In Eq. (5.50), τ_σ^{W} is the von Weizsäcker kinetic energy density, which is defined by the von Weizsäcker kinetic energy in Eq. (4.4) as

$$T_{\mathrm{W}} = \int d^3\mathbf{r}\tau^{\mathrm{W}} = \sum_\sigma \int d^3\mathbf{r}\tau_\sigma^{\mathrm{W}} = \frac{1}{8} \sum_\sigma \int d^3\mathbf{r} \frac{|\nabla\rho_\sigma(\mathbf{r})|^2}{\rho_\sigma(\mathbf{r})} . \tag{5.51}$$

The feature of the PKZB meta-GGA exchange-correlation functional is that it is the expansion of the PBE-GGA exchange-correlation functional based on a physical background. This makes it possible not only to produce more accurate exchange-correlation energies but also to estimate the contribution of the kinetic energy terms. However, the semiempirical parameters included have led to the avoidance of the use of this functional in solid state calculations.

The *TPSS meta-GGA exchange-correlation functional* (Tao et al. 2003) intends to remove the semiempirical parameters from the PKZB functional to construct a nonempirical meta-GGA functional,

$$E_{\mathrm{c}}^{\mathrm{TPSS}} = \int d^3\mathbf{r} \bar{E}_{\mathrm{c}}^{\mathrm{revPKZB}} \left[1 + d \left(\frac{\tau^{\mathrm{W}}}{\tau} \right)^3 \right] , \tag{5.52}$$

and

$$\bar{E}_{\mathrm{c}}^{\mathrm{revPKZB}} = \bar{E}_{\mathrm{c}}^{\mathrm{PBE}}[\rho_\alpha, \rho_\beta, \nabla\rho_\alpha, \nabla\rho_\beta] \left[1 + C(\zeta, \xi) \left(\frac{\tau^{\mathrm{W}}}{\tau} \right)^2 \right]$$
$$- [1 + C(\zeta, \xi)] \left(\frac{\tau^{\mathrm{W}}}{\tau} \right)^2 \sum_\sigma \frac{\rho_\sigma}{\rho} \bar{E}_{\mathrm{c}}^{\max} , \tag{5.53}$$

where

$$C(\zeta,\xi) = \frac{0.53 + 0.87\zeta^2 + 0.50\zeta^4 + 2.26\zeta^6}{\{1 + \xi^2\left[(1+\zeta)^{-4/3} + (1-\zeta)^{-4/3}\right]/2\}^4}, \tag{5.54}$$

$$\bar{E}_c^{\max} = \max\{\bar{E}_c^{\mathrm{PBE}}[\rho_\sigma, 0, \nabla\rho_\sigma, 0], \bar{E}_c^{\mathrm{PBE}}[\rho_\alpha, \rho_\beta, \nabla\rho_\alpha, \nabla\rho_\beta]\}, \tag{5.55}$$

$\xi = |\nabla\zeta|/2(3\pi^2\rho)^{1/3}$, $\zeta = (\rho_\alpha - \rho_\beta)/\rho$, $C(0,0) = 0.53$, and $d = 2.8$. Although this functional contains no semiempirical parameters, it uses six fitted fundamental constants, including four constants in $C(\zeta,\xi)$ to remove the self-interaction error (see Sect. 6.2).

For meta-GGA functionals, correlation functionals have been chiefly developed: e.g., the Filatov–Thiel 1998 (*FT98*) (Filatov and Thiel 1998) and the Krieger–Chen–Iafrate–Savin (*KCIS*) (Krieger et al. 1999) *meta-GGA correlation functionals.*

5.5 Hybrid Functionals

Hybrid functionals mix the Hartree–Fock exchange integral with GGA exchange functionals at a constant ratio, based on the concept of the adiabatic connection, which makes the Kohn–Sham energies of the independent electron model link to those of the fully interacting electron one. That is, hybrid functionals are constructed by connecting exchange functionals, which are assumed as the exchange energies of the independent electron systems, to the Hartree–Fock exchange integral, which are taken as the exchange energies for the fully interacting systems,

$$E_x = \int_0^1 d\lambda E_x^\lambda \approx E_x^{\mathrm{GGA}} + \lambda\left(E_x^{\mathrm{HF}} - E_x^{\mathrm{GGA}}\right), \tag{5.56}$$

where λ is termed the *coupling strength parameter. Equation (5.56) is different from the usual adiabatic connection, because in this particular case, the adiabatic connection is not between non-interacting and interacting systems but between non-correlated and correlated systems.* In these functionals, the electron correlation term, which is individually calculated, does not necessarily obey the concept of the adiabatic connection. Note that many publications mistakenly state that the term "hybrid functionals" is a collective designation of functionals combining the Hartree–Fock exchange integral with exchange functionals. Correctly, hybrid functionals are defined as being based on the ansatz that the exact exchange energy is situated between the GGA exchange energy functional and the Hartree–Fock exchange integral. This indicates that the mixing of the Hartree–Fock exchange integral is not a correction for exchange functionals. As hybrid functionals, various types of functionals have been developed, depending on the mixing ratios and the number of parameters, including the functionals mentioned in this section: The

B3LYP hybrid functional (Becke 1993), the PBE0 hybrid functional (Adamo and Barone 1999), and the Heyd–Scuseria–Ernzerhof (HSE) hybrid functional (Heyd et al. 2003).

The *B3LYP hybrid functional* (Becke 1993), *the first hybrid functional, is the most frequently used functional (or method) in all functionals (or all theories) in quantum chemistry calculations.* This functional uses three parameters as the mixing ratios to form the adiabatic connections between the Hartree–Fock exchange integral and the LDA exchange functional and between the LYP-GGA correlation functional and the LDA correlation functional, and to combine with the attenuated GGA term of the B88 exchange functional,

$$\begin{aligned} E_{\mathrm{xc}}^{\mathrm{B3LYP}} &= E_{\mathrm{xc}}^{\mathrm{LDA}} + a_1 \left(E_{\mathrm{x}}^{\mathrm{HF}} - E_{\mathrm{x}}^{\mathrm{LDA}} \right) + a_2 \Delta E_{\mathrm{x}}^{\mathrm{B88}} \\ &+ a_3 \left(E_{\mathrm{c}}^{\mathrm{LYP}} - E_{\mathrm{c}}^{\mathrm{VWN\text{-}LDA}} \right) \end{aligned} \tag{5.57}$$

These mixing ratios are fitted to optimize the chemical properties of the G2 benchmark set (Curtiss et al. 1991), containing several dozen atoms and small molecules. Besides the parameters in the exchange and correlation functionals, the semiempirical parameters are $a_1 = 0.2$, $a_2 = 0.72$, and $a_3 = 0.81$. For the mixing of the Hartree–Fock exchange integral, it is often interpreted that it intends to supplement the insufficient exchange interactions in the LDA functional, which causes the overestimations of binding energies, with the partial replacement of this functional with the Hartree–Fock exchange integral. Since this functional frequently gives surprisingly accurate chemical properties for small molecules, it has often been used even to justify the results of time-consuming ab initio wavefunction methods. Note, however, that *various problems have recently been reported for this functional, especially in the calculations of chemical reactions and chemical properties of large systems* (Pieniazek et al. 2008; Wheeler et al. 2009; Grimme and Korth 2007; Dreuw et al. 2003; Champagne et al. 2000).

The *PBE0 hybrid functional* (Adamo and Barone 1999) intends to physically enhance the PBE-GGA exchange-correlation functional. Based on the adiabatic approximation, using the PBE functional as the reference function, it expands this functional using the energy difference between the exchange functional and the Hartree–Fock exchange integral as the perturbation and replaces $1/4$ of the PBE exchange functional with the Hartree–Fock exchange integral to be the third-order expansion term,

$$E_{\mathrm{xc}}^{\mathrm{PBE0}} = E_{\mathrm{xc}}^{\mathrm{PBE}} + \frac{1}{4} \left(E_{\mathrm{x}}^{\mathrm{HF}} - E_{\mathrm{x}}^{\mathrm{PBE}} \right). \tag{5.58}$$

Although it is often interpreted that this functional has no parameter ("0" indicates no parameter), it contains, at least, fundamental constants in the PBE exchange-correlation functional. The advantages of this functional are its simple form, small number of parameters, and high reproducibility of chemical properties. However, this functional also has problems similar to those of the B3LYP functional.

The *HSE hybrid functional* (Heyd et al. 2003) extends the PBE exchange-correlation functional by mixing the Hartree–Fock exchange integral only for the short-range part. As explained in Sect. 6.1, it is confirmed that the lack of long-range exchange interactions is one of the major problems of GGA exchange functionals. However, it has been reported that the long-range correction for GGA exchange functionals yields much larger HOMO–LUMO gaps for semiconductors than the corresponding experimental band gaps, which solid state physicists have believed can be approximated by the HOMO–LUMO gaps. In this functional, only the short-range part of the Hartree–Fock exchange integral is mixed in the GGA exchange functionals *to make the HOMO–LUMO gaps of semiconductors approach the corresponding band gaps, which should be calculated intrinsically in excited state calculations such as time-dependent Kohn–Sham calculations*,

$$E_{\mathrm{xc}}^{\mathrm{HSE}} = aE_{\mathrm{x}}^{\mathrm{SR\text{-}HF}} + (1-a)E_{\mathrm{x}}^{\mathrm{PBE}} + E_{\mathrm{c}}^{\mathrm{PBE}}, \tag{5.59}$$

where $E_{\mathrm{x}}^{\mathrm{SR\text{-}HF}}$ is the short-range part of the Hartree–Fock exchange integral, which is separated by the error function (see Sect. 6.1). This functional uses $a = 1/4$ for the same reason as that used for the PBE0 functional, mentioned in Sect. 5.5. Of course, the main advantage of this functional is the accuracy in band calculations of semiconductors. It is actually confirmed that this functional significantly improves the properties of semiconductors, such as lattice constants, which have been seen as problems for the LDA functional, while maintaining or improving the accuracies in the calculated band energies of the LDA functional. On the other hand, the disadvantage is that the inclusion of only the short-range Hartree–Fock exchange integral cannot be easily justified. This is because the large effect of the long-range Hartree–Fock exchange integral on band energies indicates the significance of long-range exchange interactions, even for semiconductors.

Among other hybrid functionals, there are the *BHHLYP functional*, combining the B88 exchange functional and the Hartree–Fock exchange interaction in a one-to-one ratio, and the *mPW1PW91 functional* (Adamo and Barone 1998), modifying the B3LYP functional to enhance its physical validity on the basis of the PW91 exchange-correlation functional.

5.6 Semiempirical Functionals

Semiempirical functionals intend to reproduce accurate properties using as many semiempirical parameters as needed. The concept of these functionals may be based on the force fields of molecular mechanics, for example, CHARmm (Brooks et al. 1983) and Amber (Pearlman et al. 1995), which have been established to determine the structures of biomacromolecules such as folded proteins. In most cases, the development of such force fields focuses only on constructing potentials to yield highly accurate molecular structures. Likewise, semiempirical functionals have been developed to provide highly accurate properties. Note, however, that there are guidelines for developing semiempirical functionals. That is, *these functionals are represented using only dimensionless parameters,* x_σ (Eq. (5.2)) *and* z_σ (Eq. (5.45)), and *the functional forms are the expansions or modifications of conventional semiempirical functionals*. Semiempirical functionals include the B97 (Becke 1997) and the Hamprecht–Cohen–Tozer–Handy (HCTH) semiempirical functionals (Hamprecht et al. 1998), and their derivatives: the B97-series and the Mx-series semiempirical functionals.

The *B97 semiempirical functional* (Becke 1997) is the first semiempirical functional and has become the basis for conventional semiempirical functionals,

$$E_{\mathrm{xc}}^{\mathrm{B97}} = E_{\mathrm{x}}^{\mathrm{B97}} + E_{\mathrm{c}}^{\mathrm{B97}} + c_{\mathrm{x}} E_{\mathrm{x}}^{\mathrm{HF}}, \tag{5.60}$$

$$E_{\mathrm{x}}^{\mathrm{B97}} = \sum_{\sigma} \int d^3\mathbf{r} \bar{E}_{\mathrm{x}\sigma}^{\mathrm{LDA}}[\rho_\sigma] g_{\mathrm{x}\sigma}(x_\sigma^2), \tag{5.61}$$

$$g_{\mathrm{x}\sigma} = \sum_{i}^{m} C_{\mathrm{x}\sigma i} \left[\gamma_{\mathrm{x}\sigma} x_\sigma^2 \left(1 + \gamma_{\mathrm{x}\sigma} x_\sigma^2\right)^{-1} \right]^i, \tag{5.62}$$

$$\begin{aligned} E_{\mathrm{c}}^{\mathrm{B97}} &= \sum_{\sigma} \int d^3\mathbf{r} \bar{E}_{\mathrm{c}\sigma\sigma}^{\mathrm{LDA}}[\rho_\sigma] g_{\mathrm{c}\sigma\sigma}(x_\sigma^2) \\ &\quad + \int d^3\mathbf{r} \bar{E}_{\mathrm{c}\alpha\beta}^{\mathrm{LDA}}[\rho_\alpha, \rho_\beta] g_{\mathrm{c}\alpha\beta} \left(\frac{x_\alpha^2 + x_\beta^2}{2} \right), \end{aligned} \tag{5.63}$$

$$g_{\mathrm{c}\sigma\sigma} = \sum_{i}^{m} C_{\mathrm{c}\sigma\sigma i} \left[\gamma_{\mathrm{c}\sigma\sigma} x_\sigma^2 \left(1 + \gamma_{\mathrm{c}\sigma\sigma} x_\sigma^2\right)^{-1} \right]^i, \tag{5.64}$$

and

$$g_{\mathrm{c}\alpha\beta} = \sum_{i}^{m} C_{\mathrm{c}\alpha\beta i} \left[\frac{1}{2} \gamma_{\mathrm{c}\alpha\beta} (x_\alpha^2 + x_\beta^2) \left(1 + \frac{1}{2} \gamma_{\mathrm{c}\alpha\beta} (x_\alpha^2 + x_\beta^2) \right)^{-1} \right]^i, \tag{5.65}$$

where $\gamma_{\mathrm{x}\sigma} = 0.004$, $\gamma_{\mathrm{c}\sigma\sigma} = 0.2$, and $\gamma_{\mathrm{c}\alpha\beta} = 0.006$. Besides these three parameters, this functional contains ten parameters: c_{x}, $c_{\mathrm{x}\sigma i}$, $c_{\mathrm{c}\sigma\sigma i}$, and $c_{\mathrm{x}\alpha\beta i}$ for $i = 0, 1, 2$ (for the parameter values, see Becke 1997). Since this functional mixes the Hartree–Fock exchange integral at a constant ratio, it is a hybrid functional. The *HCTH semiempirical functional* (Hamprecht et al. 1998) removes c_{x} and increases the numbers of $c_{\mathrm{x}\sigma i}$, $c_{\mathrm{c}\sigma\sigma i}$, and $c_{\mathrm{x}\alpha\beta i}$ to $i = 0$ to 4 in the B97 functional. Consequently, it contains 15 parameters in total. This functional is a GGA functional, because it mixes no Hartree–Fock exchange integral. The *B97-series semiempirical functionals* are modifications of these functionals. For example, the *B97-1 functional* is different only in terms of the values of the semiempirical parameter. The *B97-D functional* (Antony and Grimme 2006) corrects the B97 functional for dispersions (see Sect. 6.3), and the *ωB97 functional* (Chai and Head-Gordon 2008) uses the long-range correction for the B97 functional (see Sect. 6.1).

The *Mx-series semiempirical functionals* (Zhao and Truhlar 2006, 2008) are hybrid meta-GGA functionals, which combine the PBE exchange functional with the B97 correlation functional and correct them using the kinetic energy density terms ("M" in Mx suggests a meta-GGA functional, and "x" is replaced by the last two digits of the year of developing the functional). For instance, the M06 functional combines the M05 functional, which is the correction of the PBE exchange functional for the kinetic energy density, and the above B97 correlation functional with the VS98 exchange-correlation functional (Sect. 5.4) and the Hartree–Fock exchange integral,

$$E_{\mathrm{x}}^{\mathrm{M06}} = \frac{X}{100}E_{\mathrm{x}}^{\mathrm{HF}} + \left(1 - \frac{X}{100}\right)E_{\mathrm{x}}^{\mathrm{M06\text{-}DFT}}, \tag{5.66}$$

$$E_{\mathrm{x}}^{\mathrm{M06\text{-}DFT}} = E_{\mathrm{x}}^{\mathrm{revM05}} + E_{\mathrm{x}}^{\mathrm{revVS98}}, \tag{5.67}$$

$$E_{\mathrm{c}}^{\mathrm{M06}} = E_{c\sigma\sigma}^{\mathrm{revB97\text{-}SIC}} + E_{c\sigma\sigma}^{\mathrm{revVS98\text{-}SIC}} + E_{c\alpha\beta}^{\mathrm{revB97}} + E_{c\alpha\beta}^{\mathrm{revVS98}}, \tag{5.68}$$

$$E_{\mathrm{x}}^{\mathrm{revM05}} = \sum_{\sigma}\int d^3\mathbf{r}\bar{E}_{\mathrm{x}\sigma}^{\mathrm{PBE}}[\rho_\sigma]f_{\mathrm{x}\sigma}(w_\sigma), \tag{5.69}$$

$$f_{\mathrm{x}\sigma}(w_\sigma) = \sum_{i}^{m} C_{\mathrm{x}\sigma i}\left(\frac{\tau_\sigma^{\mathrm{LDA}} - \tau_\sigma}{\tau_\sigma^{\mathrm{LDA}} + \tau_\sigma}\right)^i, \tag{5.70}$$

$$E_{c\sigma\sigma}^{\mathrm{M06}} = \int d^3\mathbf{r}D_\sigma\left(\bar{E}_{c\sigma\sigma}^{\mathrm{revB97\text{-}SIC}} + \bar{E}_{c\sigma\sigma}^{\mathrm{revVS98\text{-}SIC}}\right), \tag{5.71}$$

and

$$D_\sigma = 1 - \frac{x_\sigma^2}{4(z_\sigma + C_{\mathrm{F}})}, \tag{5.72}$$

where $\tau_\sigma^{\mathrm{LDA}} = C_{\mathrm{F}}\rho_\sigma^{5/3}$ and "rev" indicates the revision of the semiempirical parameters. D_σ is the self-interaction correction term. This functional contains 36 semiempirical parameters in total: $X, \gamma_{c\sigma\sigma}, \gamma_{c\alpha\beta}, C_{\mathrm{x}\sigma i}\ (i = 0\text{–}11), c_{c\sigma\sigma i}\ (i = 0\text{–}4)$, $c_{\mathrm{x}\alpha\beta i}\ (i = 0 - 4)$, $d_i\ (i = 0 - 2)$, $d_{c\sigma\sigma i}\ (i = 0 - 4)$, and $d_{c\alpha\beta i}\ (i = 0 - 4)$ (for the parameter values, see Zhao and Truhlar 2008). The *M06-2x functional* removes the VS98 exchange functional part from the M06 exchange functional and consequently uses 35 semiempirical parameters, in which $d_i\ (i = 0 - 2)$ are decreased (for the parameter values, see Zhao and Truhlar 2008). Other Mx-series functionals are the *M06-L functional*, which excludes the Hartree–Fock exchange integral (38 parameters) and the *M06-HF functional* (38 parameters), which uses the Hartree–Fock exchange integral in the exchange part. The Bose–Martin kinetic (*BMK*) *functional* (Boese and Martin 2004) is also a semiempirical functional.

As easily presumed from the concept, the above semiempirical functionals are often superior to other types of functionals in terms of the ability to reproduce chemical properties and reactions. Nevertheless, it has recently been reported that several chemical reactions are poorly reproduced (Jacquemin et al. 2010; Song et al. 2010). These functionals also depend on the benchmark sets used for determining the semiempirical parameters and the values and combinations of these parameters. Since these dependences obscure the contributory levels of constituent interactions, it would be difficult to justify the calculated results of properties that have never been experimentally observed. In the future, various types of semiempirical functionals will be developed to reproduce properties and reactions more accurately. It is, therefore, reasonable to suppose that *all conventional semiempirical functionals are inherently transitional functionals until the development of more sophisticated functionals.*

Finally, it should be emphasized that *conventional functionals have their own advantages and disadvantages and have been used to trade off these characteristics, depending on the calculated systems. It is, therefore, too optimistic to consider that functionals steadily approach the universal functional, as if climbing Jacob's ladder (Fig. 5.2) year by year, and consequently that the latest, state-of-the-art functionals are superior to the conventional ones.*

References

Adamo, C., Barone, V.: J. Chem. Phys. **108**, 664–675 (1998)
Adamo, C., Barone, V.: J. Chem. Phys. **110**, 6158–6170 (1999)
Antony, J., Grimme, S.: Phys. Chem. Chem. Phys. **8**, 5287–5293 (2006)
Becke, A.D.: Phys. Rev. A **38**, 3098–3100 (1988)
Becke, A.D.: J. Chem. Phys. **98**, 5648–5652 (1993)
Becke, A.D.: J. Chem. Phys. **107**, 8554–8560 (1997)
Boese, A.D., Martin, J.M.L.: J. Chem. Phys. **121**, 3405–3416 (2004)
Brooks, R.E., Bruccoleri, B.R., Olafson, B.D., States, D.J., Swaminathan, S., Karplus, M.: J. Comput. Chem. **4**, 187–217 (1983)
Carr, W.J.: Phys. Rev. **122**, 1437–1446 (1961)
Ceperley, D.M., Alder, B.J.: Phys. Rev. Lett. **45**, 566–569 (1980)
Chai, J.-D., Head-Gordon, M.: J. Chem. Phys. **128**, 084106(1–15) (2008)
Champagne, B., Perpéte, E.A., Jacquemin, D.: J. Phys. Chem. A **104**, 4755–4763 (2000)
Colle, R., Salvetti, O.: Theor. Chim. Acta **37**, 329–334 (1975)
Curtiss, L.A., Raghavachari, K., Trucks, G.W., Pople, J.A.: J. Chem. Phys. **94**, 7221–7230 (1991)
Dreuw, A., Weisman, J.L., Head-Gordon, M.: J. Chem. Phys. **119**, 2943–2946 (2003)
Filatov, M., Thiel, W.: Phys. Rev. A **57**, 189–199 (1998)
Gell-Mann, M., Bruecker, K.A.: Phys. Rev. **106**, 364–368 (1957)
Grimme, M., Steinmetz, S., Korth, M.: J. Org. Chem. **72**, 2118–2126 (2007)
Hamprecht, F.A., Cohen, A.J., Tozer, D.J., Handy, N.C.: J. Chem. Phys. **109**, 6264–6271 (1998)
Hedin, L., Lundqvist, B.I.: J. Phys. C Solid State Phys. **4**, 2064–2083 (1971)
Heyd, J., Scuseria, G.E., Ernzerhof, M.: J. Chem. Phys. **118**, 8207–8215 (2003)
Jacquemin, D., Perpete, E.A., Ciofini, I., Adamo, C., Valero, R., Zhao, Y., Truhlar, D.G.: J. Chem. Theory Comput. **6**, 2071–2085 (2010)
Krieger, J.B., Chen, J., Iafrate, G.J., Savin, A.: In: Gonis, A., Kioussis, N., Ciftan, M. (eds.) Electron Correlations and Materials Properties. Plenum, New York (1999)
Lee, C., Yang, W., Parr, R.G.: Phys. Rev. B **37**, 785–789 (1988)
Ma, S.K., Brueckner, K.A.: Phys. Rev. **165**, 18–31 (1968)

Negele, J.W., Vautherin, D.: Phys. Rev. C **5**, 1472–1493 (1972)
Pearlman, D.A., Case, D.A., Caldwell, J.W., Ross, W.S., Cheatham, T.E. III, DeBolt, S., Ferguson, D., Seibel, G., Kollman, P.: Comput. Phys. Commun. **91**, 1–41 (1995)
Perdew, J.P.: In: Ziesche, P., Eschrigh, H. (eds.) Electronic Structure of Solids '91. Akademie Verlag, Berlin (1991)
Perdew, J.P., Wang, Y.: Phys. Rev. B **45**, 13244–13249 (1992)
Perdew, J.P., Burke, K., Ernzerhof, M.: Phys. Rev. Lett. **77**, 3865–3868 (1996)
Perdew, J.P., Kurth, S., Zupan, A., Blaha, P.: Phys. Rev. Lett. **82**, 2544–2547 (1999)
Perdew, J.P., Ruzsinszky, A., Tao, J., Staroverov, V.N., Scuseria, G.E., Csonka, G.I.: J. Chem. Phys. **123**, 062201(1–9) (2006)
Pieniazek, S.N., Clemente, F.R., Houk, K.N.: Angew. Chem. Int. Ed. **47**, 7746–7749 (2008)
Proynov, E.I., Vela, A., Ruiz, E., Salahub, D.R.: Int. J. Quantum Chem. Symp. **29**, 61–78 (1995)
Proynov, E.I., Sirois, S., Salahub, D.R.: Int. J. Quantum Chem. Symp. **64**, 427–446 (1997)
Slater, J.C.: Phys. Rev. **81**, 385–390 (1951)
Song, J.-W., Tsuneda, T., Sato, T., Hirao, K.: Org. Lett. **12**, 1440–1443 (2010)
Tao, J., Perdew, J.P., Staroverov, V.N., Scuseria, G.E.: Phys. Rev. Lett. **91**, 146401(1–4) (2003)
Tsuneda, T., Hirao, K.: Phys. Rev. B **62**, 15527–15531 (2000)
Tsuneda, T., Suzumura, T., Hirao, K.: J. Chem. Phys. **110**, 10664–10678 (1999)
Tsuneda, T., Kamiya, M., Morinaga, N., Hirao, K.: J. Chem. Phys. **114**, 6505–6513 (2001)
van Voorhis, T., Scuseria, G.E.: J. Chem. Phys. **109**, 400–410 (1998)
Vosko, S.H., Wilk, L., Nusair, M.: Can. J. Phys. **58**, 1200–1211 (1980)
Wheeler, S.E., Moran, A., Pieniazek, S.N., Houk, K.N.: J. Phys. Chem. A **38**, 10376–10384 (2009)
Wigner, E.P.: Phys. Rev. **46**, 1002–1011 (1934)
Wigner, E.P., Seitz, F.: Phys. Rev. **43**, 804–810 (1933)
Zhang, Y., Yang, W.: Phys. Rev. Lett. **80**, 890–890 (1998)
Zhao, Y., Truhlar, D.G.: J. Chem. Phys. **125**, 194101(1–18) (2006)
Zhao, Y., Truhlar, D.G.: Theor. Chem. Acc. **120**, 215–241 (2008)

Chapter 6
Corrections for Functionals

6.1 Long-Range Correction

In this chapter, various types of corrections for incorporating physical effects neglected in exchange-correlation functionals are discussed. Since the corrections reviewed in this chapter have clear physical meanings in common with each other, they are distinctly different from empirical terms that are added simply to enhance the ability to reproduce a particular physical property. It is, therefore, possible to determine whether the corresponding physical effect is included in an exchange-correlation functional in advance and to carry out a correction for the functional if it has not already been included.

Long-range correction indicates the correction of exchange functionals for long-range electron–electron exchange interactions, which are insufficiently incorporated in conventional exchange functionals. Since exchange functionals usually depend only on the electron distribution, they essentially contain no explicit electron–electron interactions. As shown in Eq. (5.1), exchange functionals are generally represented in the form of a one-electron coordinate integral. Therefore, it is reasonable to suppose that *exchange functionals always require a long-range correction*. On the other hand, long-range exchange interactions are naturally incorporated in the Hartree–Fock exchange integral, which is an explicit two-electron coordinate integral in Eq. (2.45). The present author and coworkers have developed a long-range correction in which exchange interactions are divided into short-range and long-range parts, and then a general exchange functional and the Hartree–Fock exchange integral are adopted in the calculations of the short-range and long-range parts, respectively (Iikura et al. 2001). For the local density approximation (LDA) exchange functional (Dirac 1930), Savin suggested the formulation of the long-range correction (LC) scheme (Savin 1996). The long-range correction makes it applicable to general functionals to be useful in quantum chemistry calculations. In this correction, the two-electron operator, $1/r_{12}$, is divided by the standard error function as

T. Tsuneda, *Density Functional Theory in Quantum Chemistry*,
DOI 10.1007/978-4-431-54825-6_6,

$$\frac{1}{r_{12}} = \frac{1-\mathrm{erf}(\mu r_{12})}{r_{12}} + \frac{\mathrm{erf}(\mu r_{12})}{r_{12}}, \tag{6.1}$$

where μ is a parameter for determining the division ratio. However, it is difficult to divide exchange functionals by Eq. (6.1) in most cases, because these functionals are generally not derived from a density matrix. The long-range correction, therefore, assumes that all of the features of exchange functionals are comprised in the momentum, k_σ. Based on this assumption, the short-range part of general exchange functionals in Eq. (5.1) is derived as

$$E_{\mathrm{x}}^{\mathrm{LC(sr)}} = -\frac{1}{2}\sum_{\sigma}\int d^3\mathbf{r}\rho_\sigma^{4/3}K_\sigma\left\{1-\frac{8}{3}a_\sigma\left[\sqrt{\pi}\mathrm{erf}\left(\frac{1}{2a_\sigma}\right)+2a_\sigma\left(b_\sigma-c_\sigma\right)\right]\right\}, \tag{6.2}$$

where a_σ, b_σ, and c_σ are given as

$$a_\sigma = \frac{\mu}{2k_\sigma} = \frac{\mu K_\sigma^{1/2}}{6\sqrt{\pi}\rho_\sigma^{1/3}}, \tag{6.3}$$

$$b_\sigma = \exp\left(-\frac{1}{4a_\sigma^2}\right) - 1, \tag{6.4}$$

and

$$c_\sigma = 2a_\sigma^2 b_\sigma + \frac{1}{2}. \tag{6.5}$$

Note that momentum k_σ is correctly derived to provide the Fermi momentum, $k_{\mathrm{F}\sigma}$ in Eq. 5.8, for the LDA exchange functional in Eq. (4.3), $K_\sigma^{\mathrm{LDA}} = 3\left(3/4\pi\right)^{1/3}$, such as

$$k_\sigma = \left(\frac{9\pi}{K_\sigma}\right)^{1/2}\rho_\sigma^{1/3}. \tag{6.6}$$

Using this momentum, Eq. (6.2) is identical to the previously proposed long-range correction for the LDA functional (Savin 1996). The long-range part of the Hartree–Fock exchange integral is simply given by multiplying the standard error function by the two-electron operator in Eq. (2.45) as

$$E_{\mathrm{x}}^{\mathrm{LC(lr)}} = -\frac{1}{2}\sum_{\sigma}\sum_{i}^{n}\sum_{j}^{n}\iint d^3\mathbf{r}_1 d^3\mathbf{r}_2 \phi_{i\sigma}^*(\mathbf{r}_1)\phi_{j\sigma}^*(\mathbf{r}_1)\frac{\mathrm{erf}\left(\mu r_{12}\right)}{r_{12}}\phi_{i\sigma}(\mathbf{r}_2)\phi_{j\sigma}(\mathbf{r}_2). \tag{6.7}$$

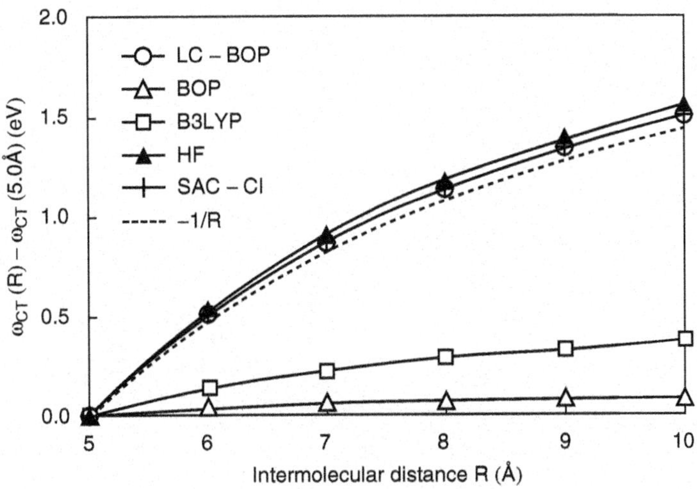

Fig. 6.1 Calculated lowest charge transfer excitation energy, ω_{CT}, of the ethylene–tetrafluoroethylene dimer with respect to the intermolecular distance, R, in eV. The excitation energy at the distance of 5 Å is set to be zero. The DFT (LC-BOP, BOP, and B3LYP) results were obtained by the time-dependent Kohn–Sham method (see Sect. 4.6), while the HF result is given by the time-dependent Hartree–Fock method. For the SAC-CI method, see Sect. 3.5. Rigorously, the excitation energy should be slightly above the curve of $-1/R$. The augmented Sadlej pVTZ basis functions are used. See Tawada et al. (2004)

In the above formulation, the only parameter μ naturally depends on the exchange functional corrected. For example, it has been confirmed that the μ value of the B88 and PBE exchange functionals is optimized as $\mu = 0.47$ for electronic ground states at around equilibrium geometries (Song et al. 2007a) and as $\mu = 0.33$ for others (Iikura et al. 2001).

The long-range correction has solved a wide variety of problems that have been reported in previous Kohn–Sham calculations. So far, *the most remarkable problems that have been resolved with the long-range correction have been the van der Waals binding energies, electronic excitation spectra, optical response properties, and orbital energies.* The reproducibilities of van der Waals binding energies and orbital energies are detailed in Sects. 6.3 and 7.9, respectively. Here, let us examine only electronic excitation spectra and optical response properties. For the electronic excitation spectra, it has been reported that the time-dependent Kohn–Sham method (see Sect. 4.6) underestimates charge transfer (Dreuw et al. 2003; Dreuw and Head-Gordon 2004) and Rydberg (Tozer and Handy 1998) excitation energies and oscillator strengths (van Gisbergen et al. 1998). It has also been reported that optical response properties such as hyperpolarizability are significantly overestimated for long-chain polyenes as the chains are lengthened in the coupled perturbed Kohn–Sham method or the finite-field method (see Sect. 4.7) (Champagne et al. 2000). As shown in Figs. 6.1 and 6.2, *the long-range correction solves the problems in the electronic excitation spectrum and optical response*

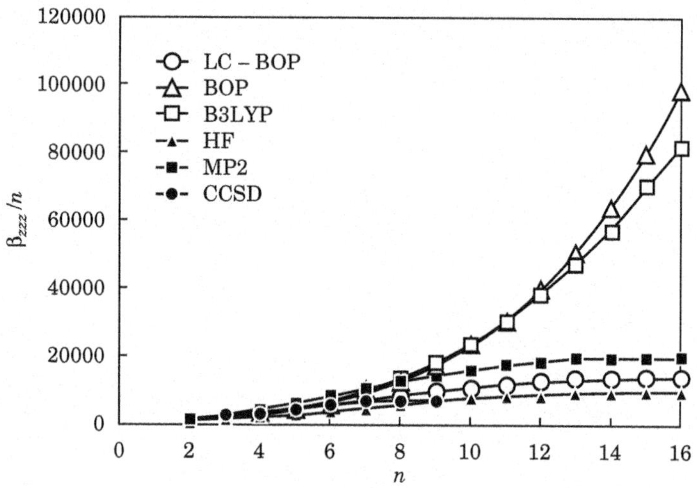

Fig. 6.2 Calculated longitudinal hyperpolarizability, β_{zzz}, of α,ω-nitro,amino-polyacetylene $O_2N(C_2H_2)_nNH_2$ with respect to the number of units n. The DFT (LC-BOP, BOP, and B3LYP) results were obtained by the coupled-perturbed Kohn–Sham method (see Sect. 4.7), the HF result was given by the coupled-perturbed Hartree–Fock method, and the ab initio results were provided by the finite-field method (see Sect. 4.7). The aug-cc-pVDZ basis functions are used. See Kamiya et al. (2005)

property calculations (Tawada et al. 2004; Kamiya et al. 2005). This indicates that *the problems are attributed to the reproducibilities of the orbital energies and exchange-correlation integral kernel.* Notably, both the time-dependent and coupled perturbed Kohn–Sham methods perform response property calculations by using the orbital energy gaps in Eqs. (4.37), (4.38), and (4.62) and the integrals containing the exchange-correlation integral kernel in matrix K of Eq. (4.41). As explained in detail in Chap. 7, without the long-range correction, the Kohn–Sham method using any conventional functional significantly underestimates either the orbital energy gaps or the exchange-correlation integral kernel. However, *since the exchange-correlation integral kernel is negative, in contrast to the positive orbital energy gaps, the calculated excitation energies often approach the correct values by the error cancellation as a whole. The most prominent example is the valence excitation energies of small molecules.* For the valence excitations of small molecules, the orbital shapes before and after excitations are similar to each other. The excitation energies seem accurately reproduced without the long-range correction by the error cancellation in such cases. However, for charge transfer and Rydberg excitations, the exchange-correlation integral kernel is very small because of the obviously different orbital shapes before and after excitations. This causes the underestimations of the orbital energy gaps to appear directly as errors in the calculated excitation energies. In the case of optical response properties, the Kohn–Sham calculations provide plausible results for small molecules due to the error cancellation. However, significant errors are obtained for long-chain molecules, which contain near-degenerate, considerably different orbitals. The oscillator strengths are also determined by the underestimated orbital energy gaps. As mentioned in Sect. 4.7, *most spectroscopic constants are response properties, similar to excitation energies and optical properties.* It is, therefore, clear that *the long-range correction plays a significant role in the calculations of spectroscopic constants.*

The high applicability of the long-range correction has been conducive to the developments of various other long-range corrected functionals. Let us briefly examine the formulations, advantages, and disadvantages of major long-range corrected functionals in this section.

The CAM-B3LYP functional (Yanai et al. 2004) is a long-range corrected hybrid functional using

$$\frac{1}{r_{12}} = \frac{1 - [\alpha + \beta \cdot \mathrm{erf}(\mu r_{12})]}{r_{12}} + \frac{\alpha + \beta \cdot \mathrm{erf}(\mu r_{12})}{r_{12}}, \tag{6.8}$$

as a substitute for Eq. (6.1) to perform the long-range correction for the B3LYP hybrid functional (see Sect. 5.5). Although various α and β have been employed, $\alpha = 0.19$ and $\beta = 0.46$ seem to be the most used. The main feature of this functional is the inclusion of the short-range Hartree–Fock exchange integral at a constant ratio, instead of the incomplete long-range exchange integral. For the correlation functional, the sum of the factor 0.19 times the VWN LDA functional and 0.81 times the LYP GGA functional is used similarly to the B3LYP functional. The CAM-B3LYP functional was developed based on *the concept that the poor atomization energies*, which have been reported as the only critical problem in the benchmark set calculations of the original long-range corrected functional, *are attributable to the uncorrected short-range exchange functional.* As a result, the calculated atomization energies are improved while maintaining the features of the original version in the benchmark set of calculations.

Contrastingly to the momentum transformation of the original long-range correction in Eq. (6.6), the LC-ωPBE functional (Vydrov et al. 2006) is derived by using the exchange hole corresponding to the Perdew–Burke–Ernzerhof exchange functional (Ernzerhof and Perdew 1998) (see Sect. 3.1) as

$$E_{\mathrm{x}}^{\mathrm{LC}\text{-}\omega\mathrm{PBE(sr)}}(\omega) = 2\pi \int d^3\mathbf{r}\rho(\mathbf{r}) \int_0^{\infty} dr_{12} \left[1 - \mathrm{erf}(\omega r_{12})\right] r_{12} h_{\mathrm{x}}^{\mathrm{PBE}}(\mathbf{r}, r_{12}), \tag{6.9}$$

where ω, which is identical to the μ of the original one, is given as $\omega = 0.4$. The advantage of this functional is *the better physical validity for the short-range exchange part than that of the original one*.

The long-range correction has also been applied to a semiempirical functional (see Sect. 5.6). The first long-range corrected semiempirical functional is the ωB97X functional (Chai and Head-Gordon 2008a) in the form,

$$E_{\mathrm{xc}}^{\omega\mathrm{B97X}} = E_{\mathrm{x}}^{\mathrm{LC(lr)}} + c_{\mathrm{x}} E_{\mathrm{x}}^{\mathrm{HF(sr)}} + E_{\mathrm{x}}^{\mathrm{B97(sr)}} + E_{\mathrm{c}}^{\mathrm{B97}}. \tag{6.10}$$

where B97 indicates the B97 semiempirical functional. For $c_{\mathrm{x}} = 0$, it is simply called the "ωB97" functional. This functional contains many semiempirical parameters: 17 parameters in ωB97X and 16 parameters in ωB97. Partly for that reason, *it frequently yields more accurate results for properties, including atomization energies, than the original ones in benchmark set calculations.* The ωB97XD functional is a van der Waals correction for the ωB97X functional, in which a parametrized classical dispersion term is combined (see Sect. 6.3).

In the LC-PR functional (Nakata and Tsuneda 2013; Nakata et al. 2010), a self-interaction correction (see Sect. 6.2) is performed for the short-range exchange part of long-range corrected functionals. Based on the regional self-interaction correction (RSIC) (Tsuneda et al. 2003), the short-range exchange part is corrected using the pseudospectral (PS) technique (Orszag 1972), such as

$$E_{\mathrm{x}}^{\mathrm{PR}} = [1 - f_{\mathrm{RS}}(t)]\, E_{\mathrm{x}}^{\mathrm{DF}} + f_{\mathrm{RS}}(t) E_{\mathrm{x}}^{\mathrm{SI}}, \tag{6.11}$$

$$E_{\mathrm{x}}^{\mathrm{SI}} = -\frac{1}{4} \sum_{\mu\nu\lambda\kappa} \int d^3\mathbf{r} P_{\mu\nu} P_{\lambda\kappa} \chi_{\nu}^{*}(\mathbf{r}) \chi_{\lambda}(\mathbf{r}) \int d^3\mathbf{r}' \frac{\chi_{\kappa}^{*}(\mathbf{r}')\chi_{\mu}(\mathbf{r}')}{|\mathbf{r}' - \mathbf{r}|}. \tag{6.12}$$

In Eq. (6.11), f_{RS} is a region-separation function for clipping the self-interaction regions (see Sect. 6.2). *This functional drastically improves the calculated core excitation energies of long-range corrected functionals in TDKS calculations*, while maintaining the accuracy in the core ionization energies, and valence, Rydberg, and charge transfer excitation energies (Nakata et al. 2010). It is also reported that *this functional provides accurate orbital energies simultaneously for core and valence orbitals* (see Sect. 7.9) (Nakata and Tsuneda 2013).

Other long-range corrected functionals have been developed: e.g., the *LCgau* (Song et al. 2007b), Mori-Sanchez–Cohen–Yang (*MCY*) (Cohen et al. 2007), and Baer–Neuhauser–Livshits (*BNL*) (Livshits and Baer 2007) functionals. In most cases, *long-range corrected functionals have common features, especially in the high reproducibility of van der Waals bonds, electronic spectra, optical response properties, and valence orbital energies.*

6.2 Self-interaction Correction

The most familiar correction for functionals may be the *self-interaction correction*, which removes the *self-interaction error* of exchange functionals. In density functional theory, *the self-interaction error indicates Coulomb self-interactions, which should cancel out with the exchange self-interactions but remain due to the use of exchange functionals* as a substitute for the Hartree–Fock exchange integral in the exchange part of the Kohn–Sham equation,

$$\Delta E^{\mathrm{SIE}} = \sum_{i}^{n} \left(J_{ii} + E_{\mathrm{xc}}[\rho_i] \right). \tag{6.13}$$

The self-interaction correction was first developed for atoms by Hartree in the early days of quantum mechanics (Hartree 1928). Then, Fermi and Amaldi suggested a self-interaction correction for the Thomas-Fermi theory (see Sect. 4.1) as a correction for the Coulomb interactions (Fermi and Amaldi 1934). The well-known form of this correction is that for the Coulomb potential,

$$V_J^{\mathrm{FA}}[\rho] = 2\left(1 - \frac{1}{n}\right)\sum_j^n \hat{J}_j[\rho]. \tag{6.14}$$

In this correction, assuming homogeneously distributed electrons like those in a uniform electron gas, the self-interaction error is removed by eliminating the Coulomb potential of one electron. Although this potential correction is based on a quite rough assumption, it is used even in chemistry, for example, to derive an exchange-correlation potential from the electron density in the ZMP method (see Sect. 4.5).

So far, the most frequently used is the *Perdew–Zunger self-interaction correction* (Perdew and Zunger 1981), which simply removes the self-interaction errors from total electronic energies,

$$E = E^{\mathrm{KS}} - \sum_i^n \left(J_{ii} + E_{\mathrm{xc}}[\rho_i]\right). \tag{6.15}$$

The self-interaction errors are also eliminated from the exchange potential,

$$V_{\mathrm{x}}(\mathbf{r}) = \frac{\delta E_{\mathrm{x}}}{\delta \rho} - \sum_i^n \left(\hat{J}_i + \frac{\delta E_{\mathrm{x}}}{\delta \rho}[\rho_i(\mathbf{r})]\right). \tag{6.16}$$

Note that this correction has the problem that the Kohn–Sham equation is not invariant for the unitary transformation of occupied orbitals, even after the correction, differently from the Hartree–Fock equation. In the Hartree–Fock equation, the variations of the Coulomb self-interaction energy and its potential for the unitary transformations of occupied orbitals cancel out with those of the exchange self-interaction, while these are not compensated, even after the correction in the Kohn–Sham equation. Therefore, the effect of the self-interaction correction depends on the difference in occupied orbitals before and after the unitary transformation. For removing this difference, it is usual to localize the orbitals before the self-interaction correction (Johnson et al. 1994). Note, however, that there are various types of orbital localization methods, and the effect of the self-interaction correction inevitably depends on them. Combining with the optimized effective potential (OEP) method (see Sect. 7.5) may be one of the most efficient ways to solve this problem. This combination enables us to consistently obtain localized potentials with no self-interaction error.

Often, self-interaction errors are also included in correlation functionals. The self-interaction errors in correlation functionals can be revealed by calculating the correlation energies of one-electron systems, which essentially must be zero. For example, LDA correlation functionals (see Sect. 5.3) generally contain parallel-spin electron correlation terms, which give correlation energies for one-electron systems, i.e., self-interaction errors. Since many GGA correlation functionals contain LDA correlation functionals, these functionals also include self-interaction errors with no self-interaction correction. The LYP correlation functional in Eq. (5.28) produces electron correlations for one-electron systems. However, a prefactor is multiplied to this functional in order to yield no electron correlation for one-electron systems. The OP correlation functional in Eq. (5.30) has no self-interaction error, at least explicitly, because it contains only opposite-spin electron correlations. Note, however, that since an exchange functional term is involved in the OP functional, it is possible for the OP functional to have self-interaction errors implicitly, although these errors would be negligible. As mentioned above, correlation functionals usually contain self-interaction errors. Nevertheless, the self-interaction correction for correlation functionals has rarely been used, because the errors are assumed to be much smaller than those in exchange functionals.

The self-interaction correction has chiefly been applied to solid state physics calculations. In particular, many band calculations have used self-interaction corrections to improve the discrepancies in band gaps, i.e., excitation energies, and orbital energy gaps in Kohn–Sham calculations using pure functionals. This is because the self-interaction correction tends to enlarge the exchange energies of occupied orbitals. Actually, it has been reported that the self-interaction correction clearly leads to underestimated band gaps of LDA functional, close to the experimental values (Svane and Gunnarsson 1990). However, recent calculations using atom-centered basis sets for core orbitals have shown that *the band gaps of semiconductors are well reproduced by pure functionals without self-interaction correction* (Gerber et al. 2007) and that *this correction instead leads to the overestimation of band gaps.* It has also been found that *the band gaps of insulators are overestimated by using the self-interaction correction* (Arai and Fujiwara 1995). In fact, it is accepted that *the lack of long-range exchange interactions causes the underestimation of the band gaps of insulators at least* (see Sect. 7.9) (Gerber et al. 2007).

Self-interaction errors are more serious for core orbitals than for valence orbitals. For σ-spin self-interacting electrons, which have no electron-electron interaction, the density matrix is represented as

$$P_\sigma(\mathbf{r}_1, \mathbf{r}_2) = \rho_\sigma^{1/2}(\mathbf{r}_1)\rho_\sigma^{1/2}(\mathbf{r}_2). \tag{6.17}$$

Using this density matrix, the kinetic energy density, τ, becomes the Weizsäcker one, τ^{W}, in Eq. (4.4) (Dreizler and Gross 1990), such as

$$\tau = -\frac{1}{2}\sum_\sigma \nabla^2 P_\sigma(\mathbf{r}_1, \mathbf{r}_2)\bigg|_{\mathbf{r}_1=\mathbf{r}_2} \to \tau^{\mathrm{W}} = \sum_\sigma \frac{1}{8}\frac{|\nabla\rho_\sigma|^2}{\rho_\sigma}. \tag{6.18}$$

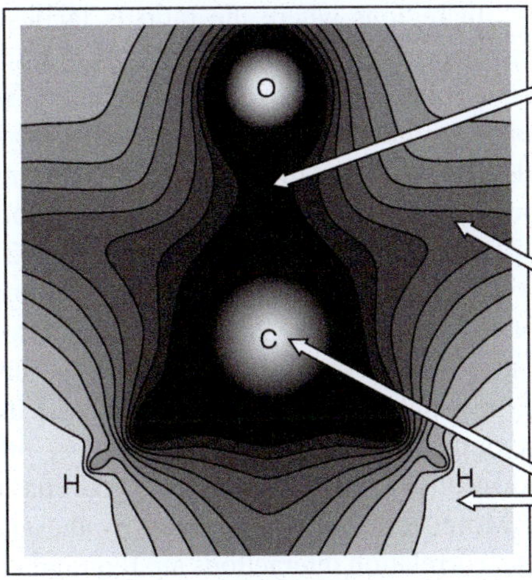

Fig. 6.3 Ratio of the Weizsäcker kinetic energy density to the total one, $\tau_W/\tau_{\text{total}}$ in Eq. (6.18), in the formaldehyde molecule with three regions of electrons

The exchange energy density meets *the far-from-nucleus (long-range) asymptotic behavior condition* in Eq. (8.26) of Chap. 8 on the basis of this density matrix, such as

$$\bar{E}_{x\sigma}(\mathbf{r}) = -\frac{1}{2}\int d^3\mathbf{r}' \frac{|P_\sigma(\mathbf{r},\mathbf{r}')|^2}{|\mathbf{r}-\mathbf{r}'|} \xrightarrow{r\to\infty} -\frac{\rho_\sigma(\mathbf{r})}{2r}. \tag{6.19}$$

Furthermore, the parallel-spin correlation energy density is assumed to be zero for self-interacting electrons, because the pair density matrix (the diagonal term of the second-order density matrix) using this density matrix becomes zero for parallel-spin electrons,

$$P_{2\sigma\sigma}(\mathbf{r}_1,\mathbf{r}_2) = \frac{1}{2}\left[\rho_\sigma(\mathbf{r}_1)\rho_\sigma(\mathbf{r}_2) - |P_\sigma(\mathbf{r}_1,\mathbf{r}_2)|^2\right] = 0 \Longrightarrow \bar{E}_{c\sigma\sigma} = 0. \tag{6.20}$$

Using the relation in Eq. (6.18), *self-interaction regions*, where electrons have no electron-electron interaction, can be identified (Tsuneda et al. 2001). That is, *those regions where the Weizsäcker kinetic energy density is close to the total one, are self-interaction regions.* Note that the Weizsäcker kinetic energy density is the minimum of the total one. Therefore, the ratio of the Weizsäcker kinetic energy density to the total is suitable to specify the self-interaction regions in molecules. Figure 6.3 illustrates this ratio in the formaldehyde molecule (Tsuneda et al. 2003). As shown in this figure, the self-interaction regions are localized only in the regions near the atomic nuclei and around the hydrogen atoms opposite to the chemical bonds. That is, the *self-interacting electrons are present only in the core orbitals and the nonbonding regions of valence 1s orbitals.* Other regions are classified into those around bonds, where the ratio is far less than 1, and the closely situated regions, where the ratio is not so far from 1. In regions where the ratio is far less than 1, the electrons are assumed to be in motion, like the free electrons in transition metals. Therefore, these regions are interpreted as *free-electron regions* (Tsuneda et al. 2001). The exchange-correlation potential in these regions is considered to be well described as a density functional. Actually, since the electron distributions vary slowly in these regions, the kinetic, exchange, and correlation energy components can be described as the expansions of dimensionless parameters in Eq. (5.2), and consequently, it is revealed that *the kinetic, exchange, and correlation energies have a transversing physical connection* (Tsuneda et al. 2001). It is appropriate to interpret the remaining intermediate regions, where the ratio is not so far from 1, as *long-range interaction regions*, in which widely separated electrons interact with each other. In Fig. 6.3, these regions are located far from the core orbitals. This is consistent with the calculated results that the long-range correction hardly affects core orbitals (see Sect. 6.1). Moreover, this is supported by the analysis result that the long-range exchange and correlation interactions are required besides the self-interaction correction to describe chemical bonds accurately (Gräfenstein et al. 2004). The above overview, therefore, suggests that *both the long-range and self-interaction corrections are needed in exchange-correlation functionals* (Nakata et al. 2010).

6.3 van der Waals Correction

The *van der Waals interaction* is one of the most significant types of electron correlations, even though it has been neglected in the development of most correlation functionals. By definition, the van der Waals interaction is a collective term that includes dipole–dipole, dipole-induced dipole, and dispersion interactions (Israelachvili 1992). *The dipole–dipole interaction* is the electrostatic interaction between permanent dipoles in polar systems. For the interactions between systems A and B, the corresponding potential is given classically as

$$V_{\mu\text{-}\mu}(\mathbf{r}) = -\frac{\mu_{\rm A}\mu_{\rm B}}{{R_{\rm AB}}^3}, \tag{6.21}$$

where $\mu_{\rm X}$ is the permanent dipole of system X and $R_{\rm AB}$ is the distance between systems A and B. In the Kohn–Sham equation, this interaction is contained as a part of the Coulomb interactions. *The dipole-induced dipole interaction* is the interaction between polar and nonpolar systems. Assuming the permanent dipole moment of polar system A as $\mu_{\rm A}$ and the polarizability, the linear response for an electric field producing an induced dipole moment of nonpolar system B as $\alpha_{\rm B}$, the classical potential of this interaction is given as (Israelachvili 1992)

$$V_{\mu\text{-}\alpha}(\mathbf{r}) = -\frac{\mu_{\rm A}^2\alpha_{\rm B}}{{R_{\rm AB}}^6}. \tag{6.22}$$

Since this interaction is very weak, it causes the low solubility of polar molecules in nonpolar solvents. This interaction is also incorporated in Kohn–Sham SCF calculations. *The dispersion interaction* is a universal interaction, which acts even between bodies with neither charge nor multipole moment. As a classical expression, the potential function between two heterogeneous bodies, which London developed using perturbation theory, is used (London 1930),

$$V_{\mathrm{disp}}^{\mathrm{London}}(\mathbf{r}) = -\frac{3}{2}\frac{\alpha_{\mathrm{A}}\alpha_{\mathrm{B}}}{R_{\mathrm{AB}}{}^{6}}\frac{I_{\mathrm{A}}I_{\mathrm{B}}}{I_{\mathrm{A}}+I_{\mathrm{B}}}, \tag{6.23}$$

where I_{X} is the ionization potential of partial system X. This dispersion interaction is interpreted as the interaction between an instantaneous dipole moment, which is caused by a fluctuation of the electron distribution, and an induced dipole moment, due to the electric field formed by the instantaneous dipole moment. That is, two spatially separated electron distributions fluctuate around their equilibrium distributions by electron correlation to produce interactions between the two bodies. Therefore, the dispersion interaction is a pure electron correlation between two bodies, which cannot be incorporated in the one-body mean-field approximation, and it is a long-range correlation explicitly acting between distant electrons. *This dispersion interaction is the only van der Waals interaction that is not incorporated in Kohn–Sham calculations using conventional correlation functionals.*

Even though the dispersion interaction should be included in correlation functionals, it has usually not been taken into consideration. Most conventional GGA correlation functionals have been developed by density gradient corrections for the LDA correlation functional or by incorporating the dynamical correlation coming from the correlation cusp. Since these functionals contain only short-range correlation resulting from correlation holes, long-range correlations, including dispersion interactions, are neglected in these functionals. Naturally, Kohn–Sham calculations using these GGA correlation functionals have almost always failed, with no dispersion interaction, to give van der Waals bonds even qualitatively. It is, therefore, reasonable to consider that dispersion interactions should be explicitly supplemented in conventional correlation functionals. So far, various types of dispersion corrections have been suggested. These dispersion corrections are generally classified into five types: classical dispersion corrections, combinations with perturbation theories, linear-response theories, van der Waals (dispersion) functionals, and semiempirical dispersion-corrected functionals. In addition to these dispersion corrections, long-range exchange interactions and correlation functionals are also significant in calculating van der Waals bonds.

The simplest dispersion correction may be the empirical correction for the Kohn–Sham energy using the London classical interatomic dispersion energy,

$$E_{\mathrm{disp}}^{\mathrm{London}} = -\sum_{\mathrm{A>B}}\frac{C_{6}^{\mathrm{AB}}}{R_{\mathrm{AB}}{}^{6}}f_{\mathrm{damp}}(R_{\mathrm{AB}}), \tag{6.24}$$

where A and B are usually the labels of atoms and C_6^{AB} is a parameterized interatomic dispersion coefficient, and f_{damp} is a damping function for cutting off unnecessary short-range interactions. The special feature of this correction method is the much shorter computational time required in dispersion calculations compared to those of other dispersion corrections. Since this classical correction also accurately reproduces experimental results with well-calibrated dispersion coefficients, it is the most widely used dispersion correction in classical molecular dynamics (MD) simulations. *It is, however, difficult to apply this correction to the calculations of new systems about which little is known, due to the empirically parameterized dispersion coefficient* and it has been reported that *this correction gives much different results, depending on the exchange-correlation functionals combined.* Therefore, the right-hand side of Eq. (6.24) is usually multiplied by adjusted parameters that are dependent on the exchange-correlation functional that is combined, as seen in DFT-D functionals, mentioned later. However, these methods are too empirical to be appropriate for a wide variety of systems. To solve this problem, Becke developed the exchange-hole dipole moment *(XDM) method* (Becke and Johnson 2005a,b), in which the C_6^{AB} coefficient in Eq. (6.24) is calculated by

$$C_6^{AB} = \frac{\langle d_X^2 \rangle_A \langle d_X^2 \rangle_B \alpha_A \alpha_B}{\langle d_X^2 \rangle_A \alpha_B + \langle d_X^2 \rangle_B \alpha_A}, \tag{6.25}$$

$$\langle d_X^2 \rangle_A = \sum_{\sigma} \int d^3\mathbf{r} \rho_\sigma(\mathbf{r}) d_{X\sigma}^2(\mathbf{r}), \tag{6.26}$$

and

$$\mathbf{d}_{X\sigma}(\mathbf{r}) = \left\{ \frac{1}{\rho_\sigma(\mathbf{r})} \sum_{ij} \left[\int d^3\mathbf{r}' \mathbf{r}' \phi_{i\sigma}(\mathbf{r}') \phi_{j\sigma}(\mathbf{r}') \right] \phi_{i\sigma}(\mathbf{r}) \phi_{j\sigma}(\mathbf{r}) \right\} - \mathbf{r}. \tag{6.27}$$

Although this method is also empirical, it has a physical meaning, at least in the dispersion coefficient, and gives more accurate dispersion interactions than those of the London classical dispersion energy.

Perturbation theories such as the MP2 method (McWeeny 1992) (see Sect. 3.2) have been appreciated as ab initio wavefunction theories reproducing dispersion interactions with relatively short computational times. Therefore, *dispersion interactions can be incorporated in the Kohn–Sham method by combining with such perturbation theories, in principle.* One of the methods based on this concept is the *DFT symmetry-adapted perturbation theory (DFT–SAPT)*, which uses Kohn–Sham orbitals to calculate the perturbation energies (Williams and Chabalowski 2001). In contrast to ab initio SAPT, in which both intermolecular and intramolecular electron correlations are calculated, only intermolecular electron correlations are calculated as a dispersion correction for the Kohn–Sham method in DFT–SAPT. Consequently, this drastically reduces the computational cost, typically by one or two orders of magnitude, compared to an ab initio SAPT calculation, with similar accuracies.

Although DFT–SAPT is a promising dispersion calculation method for clearly separated systems, *it cannot reproduce intramolecular dispersion interactions.* Moreover, despite the drastically reduced computational time, *DFT–SAPT calculations need much more computational time than those of Kohn–Sham calculations, even for the lowest order DFT–SAPT2.* Consequently, this approach is applicable to systems containing only up to several dozens of atoms with currently available computer performance. As another method combining DFT with perturbation theories, there are *double-hybrid functionals*, which mix perturbation energies in correlation functionals at a constant ratio (Schwabe and Grimme 2007). That is, these functionals extend hybrid functionals by mixing correlation functionals with MP2 perturbation energies, for example,

$$E_{\mathrm{xc}} = (1 - a_{\mathrm{x}})E_{\mathrm{x}} + a_x E_{\mathrm{HF}} + (1 - a_{\mathrm{c}})E_{\mathrm{c}} + a_{\mathrm{c}} E_{\mathrm{MP2}}, \tag{6.28}$$

where E_{MP2} is the MP2 electron correlation energy. The *B2PLYP functional* (Grimme 2006) is one of these double-hybrid functionals. Although this method gives middle-range electron correlations, it is not appropriate for dispersion calculations because of the incomplete dispersion interactions incorporated.

By using linear-response theories, dispersion interactions can be calculated directly in the framework of the Kohn–Sham method. The *adiabatic connection/fluctuation-dissipation theorem (AC/FDT) method* is a linear-response theory *for exactly calculating dispersion interactions within the framework of the Kohn–Sham method* (Langreth and Perdew 1975). In this AC/FDT method, electron correlation is calculated as the energy response quantity for the spontaneous fluctuations of electronic motions coming from the perturbation of the interelectronic interactions, as follows:

$$\begin{aligned} E_{\mathrm{c}} = -\int_0^1 d\lambda \iint d^3\mathbf{r} d^3\mathbf{r}' \frac{1}{|\mathbf{r} - \mathbf{r}'|} \\ \times \left[\frac{1}{2\pi} \int_0^{\infty} du \left\{ \chi_\lambda \left(\mathbf{r}, \mathbf{r}', iu\right) - \chi_0 \left(\mathbf{r}, \mathbf{r}', iu\right) \right\} \right], \end{aligned} \tag{6.29}$$

where $\mathbf{r}$ and $\mathbf{r}'$ are the position vectors of electrons. In this equation, χ_λ and χ_0 are density response functions for interacting and independent electrons, respectively, and these are obtained by solving the Dyson equation:

$$\begin{aligned} \chi_\lambda \left(\mathbf{r}, \mathbf{r}', \omega\right) &= \chi_0 \left(\mathbf{r}, \mathbf{r}', \omega\right) \\ &\quad + \iint d^3\mathbf{r}_1 d^3\mathbf{r}_2 \chi_0 \left(\mathbf{r}, \mathbf{r}_1, \omega\right) \\ &\quad \times \left\{ \frac{\lambda}{|\mathbf{r}_1 - \mathbf{r}_2|} + f_{\mathrm{xc}}^{\lambda} \left(\mathbf{r}_1, \mathbf{r}_2, \omega\right) \right\} \chi_\lambda \left(\mathbf{r}_2, \mathbf{r}', \omega\right), \end{aligned} \tag{6.30}$$

where f_{xc}^{λ} is the exchange-correlation integral kernel for interacting systems. Since the electron correlation in Eq. (6.29) contains dispersion interactions as the long-range correlation, the long-range part of this correlation energy is often used as a dispersion correction, which is termed a "*RPAx*" dispersion correction (Zhu et al. 2010). Analogously to the TDKS method (see Sect. 4.6), the correlation energy is calculated by solving the TDKS matrix equation. This AC/FDT correlation energy has also been found to contain a certain amount of nondynamical correlation effects (Bleiziffer et al. 2012). Although this AC/FDT method is clearly the most superior dispersion correction from a physical point of view, *it requires an enormous amount of computational time*, at least three orders of magnitude greater than the time needed in Kohn–Sham calculations, unless drastic approximations were to be adopted.

Van der Waals (dispersion) functionals have been developed *to reduce the enormous computational time required in the AC/FDT method while maintaining accuracy and ease of use, as in the London classical potential.* Lundqvist and coworkers proposed a dispersion functional, called the Andersson–Langreth–Lundqvist (ALL) functional by using a local density approximation for the electron density response function of the AC/FDT method (Andersson et al. 1996),

$$E_{\mathrm{disp}}^{\mathrm{ALL}}[\rho] = -\frac{6}{4\pi^{3/2}} \int_{V_1} d^3\mathbf{r}_1 \int_{V_2} d^3\mathbf{r}_2 \frac{\sqrt{\rho(\mathbf{r}_1)}\sqrt{\rho(\mathbf{r}_1)}}{\sqrt{\rho(\mathbf{r}_1)} + \sqrt{\rho(\mathbf{r}_1)}} \frac{1}{r_{12}^6}. \tag{6.31}$$

As almost the same functional, Dobson and coworkers independently suggested a dispersion functional based on the *local response approximation* in the same year (Dobson and Dinte 1996). Note that explicit numerical two-electron integrals, which usually require long computational times, are included in the formulations of these functionals. However, in actual calculations, *the computational time is usually less than that of the Kohn–Sham calculation*, because spatial regions of small momentum variations and core regions can be neglected in the integral calculations. On the other hand, these functionals are applicable only to two-body systems having well-separated electron distributions and *need damping functions f_{damp}* shown in Eq. (6.24) *for short-range electron–electron distances*. To solve this problem, many researchers have attempted *to develop dispersion functionals applicable to regions of overlapping electron distributions*. Lundqvist and coworkers also proposed a dispersion functional available for such electronic regions (Dion et al. 2004). This functional has a complicated form using the $\phi(\mathbf{r}_1, \mathbf{r}_2)$ function, containing the spatial coordinates of two electrons and the electron density and its gradient at these coordinates,

$$E_{\mathrm{disp}}^{\mathrm{DRSLL}}[\rho] = \int d^3\mathbf{r}_1 \int d^3\mathbf{r}_2 \rho(\mathbf{r}_1)\phi(\mathbf{r}_1, \mathbf{r}_2)\rho(\mathbf{r}_2), \tag{6.32}$$

$$\phi(\mathbf{r}_1, \mathbf{r}_2) = \frac{2}{\pi^2} \int_0^{\infty} daa^2 \int dbb^2 W(a, b) T\left(\nu_1(a), \nu_1(b), \nu_2(a), \nu_2(b)\right), \tag{6.33}$$

$$T(w,x,y,z) = \frac{1}{2}\left[\frac{1}{w+x}+\frac{1}{y+x}\right]\left[\frac{1}{(w+y)(x+z)}+\frac{1}{(w+z)(y+x)}\right], \tag{6.34}$$

and

$$W(a,b) = \frac{2}{a^3b^2}\left[(3-a^2)b\cos b\sin a + (3-b^2)a\cos a\sin b \right.$$
$$\left. +(a^2+b^2-3)\sin a\sin b - 3ab\cos a\cos b\right], \tag{6.35}$$

where

$$\nu_i(y) = y^2/2\left(1-\exp(-4\pi y^2/9d_i^2)\right), \tag{6.36}$$

and

$$d_i = r_{12}q_0(\mathbf{r}_i). \tag{6.37}$$

In Eq. (6.37), q_0 is given using the Fermi momentum, $k_F = (3\pi^2\rho)^{1/3}$, as

$$q_0(\mathbf{r}) = k_F(\mathbf{r})\left[1+0.09434\left(\frac{\nabla\rho(\mathbf{r})}{2k_F(\mathbf{r})\rho(\mathbf{r})}\right)^2\right]. \tag{6.38}$$

This functional requires no damping function, because it naturally approaches zero for short electron–electron distances. Therefore, this functional can reproduce intramolecular dispersion interactions. This dispersion functional is used, for example, in the *vdW-DF method*, (Dion et al. 2004) combining with the revPBE functional or other GGA functionals. As other interesting dispersion functionals, Vydrov–van Voorhis 2009 (*VV09*) *functional* (Vydrov and van Voorhis 2009), which uses the dielectric model smoothing the real-space cutoff, and local response dispersion (*LRD*) *functional* (Sato and Nakai 2009), which combines the local response approximation functional of Dobson and coworkers (Dobson and Dinte 1996) with the real-space cutoff (Vydrov and van Voorhis 2009), are also suggested. Combining with long-range corrected DFT, this LRD functional has succeeded to reproduce various kind of reaction energy diagrams and photochemistries, which have been poorly given by conventional DFTs.

Finally, the *semiempirical dispersion-corrected functionals* are a modification of conventional semiempirical functionals (see Sect. 5.6) for dispersion effects. *DFT-D functionals* such as the BLYP-D, B3LYP-D, and B97-D functionals (Antony and Grimme 2006) and several Mx-series functionals such as the M05-2x and M06-2x functionals (Zhao and Truhlar 2008) are included in these semiempirical dispersion-corrected functionals. In the DFT-D functionals, there are three versions, DFT-D1, DFT-D2, and DFT-D3, based on the level of dispersion corrections. For a deep understanding of the dispersion corrections, it is interesting to examine the highest level DFT-D3 functional (Grimme et al. 2010),

$$E^{\text{DFT-D3}} = E^{\text{KS}} + E^{(2)}_{\text{disp}} + E^{(3)}_{\text{disp}} \tag{6.39}$$

$$E^{(2)}_{\text{disp}} = -\sum_{\text{A}>\text{B}} \sum_{n=6,8,10,\cdots} s_n \frac{C_n^{\text{AB}}}{R_{\text{AB}}{}^n} f^n_{\text{damp}}(R_{\text{AB}}), \tag{6.40}$$

$$E^{(3)}_{\text{disp}} = -\sum_{\text{A}>\text{B}>\text{C}} \frac{C_9^{\text{ABC}}(3\cos\theta_a \cos\theta_b \cos\theta_c + 1)}{(R_{\text{AB}}R_{\text{BC}}R_{\text{CA}})^3} f^{(3)}_{\text{damp}}(\bar{R}_{\text{ABC}}), \tag{6.41}$$

where A, B, and C are atomic labels, θ_a, θ_b, and θ_c are the internal angles of the ABC triangles, and $\bar{R}_{\text{ABC}}$ are geometrically averaged radii. The damping functions are given by

$$f^n_{\text{damp}}(R_{\text{AB}}) = \frac{1}{1 + 6\left(R_{\text{AB}}/(s_{r,n} R_0^{\text{AB}})\right)^{-\alpha_n}}, \tag{6.42}$$

and

$$f^{(3)}_{\text{damp}}(\bar{R}_{\text{ABC}}) = \frac{1}{1 + 6\left(\bar{R}_{\text{ABC}}/(4R_0^{\text{ABC}}/3)\right)^{-16}}, \tag{6.43}$$

All of the remaining parameters are semiempirical: R_0^{AB} and R_0^{ABC} are cutoff radii adjusted to each pair and trio of atoms. Coefficients s_n $(n = 8, 10, \cdots)$ are fitted in benchmark calculations depending on the functionals combined, while s_6 is unity or an adjusted value less than unity. For dispersion coefficients, C_n^{AB} and C_9^{ABC}, the TDKS method and recursion relations are used to determine the values for each atomic pair and trio. The lowest-order C_6^{AB} is expressed in the Casimir–Polder formula (Casimir and Polder 1948),

$$C_6^{\text{AB}} = \frac{3}{\pi} \int_0^\infty d\omega \alpha^{\text{A}}(i\omega)\alpha^{\text{B}}(i\omega), \tag{6.44}$$

where $\alpha^{\text{A}}(i\omega)$ is the averaged dipole polarizability of atom A at an imaginary frequency. Other C_n^{AB} coefficients are derived by the recursion relations (Starkschall and Gordon 1972) as

$$C_8^{\text{AB}} = 3C_6^{\text{AB}}\sqrt{Q_{\text{A}}Q_{\text{B}}}, \tag{6.45}$$

$$C_{10}^{\text{AB}} = \frac{49}{40}\frac{(C_8^{\text{AB}})^2}{C_6^{\text{AB}}}, \tag{6.46}$$

and

$$C_{n+4}^{\text{AB}} = C_{n-2}^{\text{AB}} \left(\frac{C_{n+2}^{\text{AB}}}{C_n^{\text{AB}}}\right)^3, \tag{6.47}$$

with

$$Q_{\mathrm{A}} = \sqrt{Z_{\mathrm{A}}} \frac{\langle r^4 \rangle_{\mathrm{A}}}{\langle r^2 \rangle_{\mathrm{A}}}, \tag{6.48}$$

where Z_{A} is the nuclear charge of atom A and $\langle r^4 \rangle_{\mathrm{A}}$ and $\langle r^2 \rangle_{\mathrm{A}}$ are the expectation values derived from the electron density of atom A. The remaining triple–dipole C_9^{ABC} coefficient is given by

$$C_9^{\mathrm{ABC}} = \frac{3}{\pi} \int_0^{\infty} d\omega \alpha^{\mathrm{A}}(i\omega)\alpha^{\mathrm{B}}(i\omega)\alpha^{\mathrm{C}}(i\omega) \approx -\sqrt{C_6^{\mathrm{AB}} C_6^{\mathrm{BC}} C_6^{\mathrm{CA}}}. \tag{6.49}$$

These coefficients are determined by the TDKS calculation for each atomic pair and trio. Since these values are fixed after one TDKS calculation, these dispersion calculations are not a bottleneck in the DFT-D3 calculations. For DFT-D3 functionals, the BLYP-D3 and B2PLYP-D3 functionals have recently been suggested (Grimme et al. 2010). In the Mx-series and other semiempirical dispersion-corrected functionals, dispersion interactions are incorporated in a similar manner, although only the C_6^{AB} term is usually retained. This type of dispersion correction is clearly efficient, because *dispersion interactions are easily incorporated with high accuracy by using only functionals.* However, it has been reported that *the calculated results depend on the parameters used and the R^{-6} decay of the dispersion interaction cannot be reproduced.*

Although only dispersion corrections have so far been presented in this section, we should note that *repulsions balanced with dispersion attractions are equivalently significant in dispersion calculations.* As seen in the r^{-12} repulsion term of the Lennard-Jones potential (Lennard-Jones 1924), the repulsions have been interpreted to come from long-range exchange interactions. Nevertheless, long-range exchange interactions have been neglected in conventional exchange functionals, similarly to dispersion interactions in conventional correlation functionals. Figure 6.4 displays dissociation potential energy curves for the Ar dimer, which are calculated using various exchange functionals (Kamiya et al. 2002). It is well known that the bond in the Ar dimer consists only of dispersion interactions. This figure clearly indicates that the dissociation potentials of van der Waals bonds are significantly affected by the exchange functionals used. In contrast, the right-hand panel of the figure shows that the long-range correction (see Sect. 6.1) clearly makes these potential energy curves approach each other. This result reveals that the differences in the potential energy curves for the GGA functionals mainly come from the lack of long-range exchange interactions. Therefore, this suggests that *long-range exchange repulsions balanced with dispersion attractions are required to give accurate dispersion bonds.*

The Kohn–Sham method, when it incorporates both long-range corrected exchange and dispersion correlation effects, can reproduce accurate dispersion bonds. Figure 6.4 also illustrates the potential energy curves of the *LC+vdW method* (Kamiya et al. 2002), in which a long-range corrected functional is combined

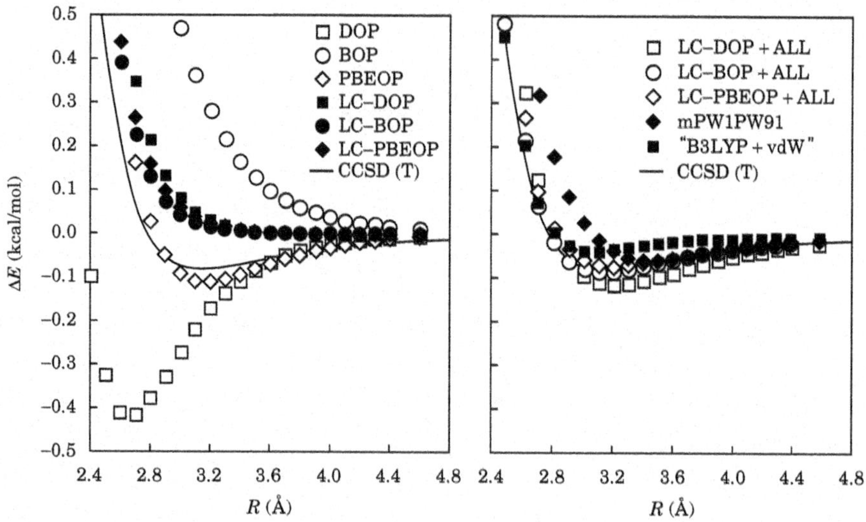

Fig. 6.4 Calculated potential energy curves of Ar_2 in terms of the inter-atomic distance: Kohn–Sham potential energy curves of pure GGA and LC-GGA functionals (*left*) and of dispersion-corrected functionals (*right*). For comparison, the curve of the CCSD(T) method (see Sect. 3.5) is also illustrated. The 6-311++G(3df, 3pd) basis functions are used. See Tsuneda and Sato (2009)

with a dispersion functional, the ALL functional in Eq. (6.31). As shown in the figure, LC functionals give very accurate potential energy curves, which are close to the results of an ab initio CCSD(T) calculation (see Sect. 3.5) (Giese et al. 2003), in contrast to those of other dispersion-corrected functionals. In addition, it has been suggested that long-range exchange interactions play a determining role in the van der Waals bond angles (Sato et al. 2007). Besides the ALL dispersion functional, LC functionals have been combined with various dispersion corrections mentioned above: e.g., the LRD dispersion functional, semiempirical AC/FDT dispersion energy functionals such as ωB97X-D (Chai and Head-Gordon 2008b), and perturbation energies such as ωB97X-2 (Chai and Head-Gordon 2009). Actually, LC functionals can be combined easily with any dispersion correction.

In Kohn–Sham calculations of dispersion interactions, the choice of correlation functional is also important. Figure 6.4 also compares the dissociation potential energy curves for the Ar dimer using the OP and LYP correlation functionals, which are both Colle–Salvetti-type correlation functionals (see Sect. 5.3). As shown in the figure, a bond is formed when using the LYP functional, while repulsive PESs are produced when using the OP functional. Since these Colle–Salvetti-type functionals essentially neglect long-range electron correlations, including dispersion interactions (see Sect. 5.3), it does not make sense that a functional using the LYP functional would provide a dispersion bond with no dispersion correction. This is attributed to the inappropriate behavior of the LYP functional. *To produce appropriate weak bonds such as van der Waals bonds, correlation functionals*

Table 6.1 Mean absolute deviations (MAD) of Kohn–Sham calculations using various types of dispersion corrections for the S22 benchmark set in ascending order in kcal/mol

Method	Type of correction(s)	MAD
ωB97X-D[a]	LC + semiempirical	0.22
BLYP-D3[b]	Semiempirical	0.23
ωB97X-2[c]	LC + perturbation	0.26
LC-BOP+LRD[d]	LC + vdW functional	0.27
B2PLYP-D3[b]	Semiempirical + perturbation	0.29
RSH+RPAx-SO2[e]	LC + AC/FDT	0.41
M06-2x[f]	Semiempirical	0.44
BLYP-D[g]	Semiempirical	0.55
B97-D[g]	Semiempirical	0.61
B3LYP-D[g]	Semiempirical	0.70
MP2/CBS[b]	Perturbation	0.78
HF+VV09[h]	vdW functional	0.89
M05-2x[f]	Semiempirical	0.90
vdW-DF(rPW86)[h]	vdW functional	1.03
rPW86+VV09[h]	vdW functional	1.20
vdW-DF(revPBE)[h]	vdW functional	1.44
vdW-DF(HF)[h]	vdW functional	2.80

The MAD of the MP2 method at the CBS limit is also displayed for comparison
[a]Chai and Head-Gordon (2008b), [b]Grimme et al. (2010), [c]Chai and Head-Gordon (2009), [d]Sato and Nakai (2009), [e]Toulouse et al. (2011), [f]Pernal et al. (2009), [g]Antony and Grimme (2006), [h]Vydrov and Van Voorhis (2010)

must obey the high-density-gradient-low-density (HDGLG) *limit condition for correlation energy* in Eq. (8.8) of Chap. 8. However, the LYP functional does not meet this condition, differently from the OP functional. The Kohn–Sham method, when using a correlation functional violating this condition, usually overstabilizes van der Waals bonds. Therefore, *a correlation functional meeting the HDGLG limit condition should be carefully chosen in Kohn–Sham calculations of dispersion interactions.*

As a benchmark set for the quantitative validation of dispersion corrections, Hobza and coworkers suggested *the S22 set* (Jurecka et al. 2006), which contains 22 weakly bonded dimers. This S22 benchmark set provides interaction energies of hydrogen-bonded, dispersion-bonded and mixed complexes, which are the calculated results of the CCSD(T) method at the complete basis set (CBS) limit (Riley et al. 2010). Due to its convenience, this benchmark set has been used not only in testing the accuracies of dispersion corrections but also in determining the adjustable parameters of semiempirical dispersion-corrected functionals. Table 6.1 displays the mean absolute deviations (MAD) of various dispersion-corrected DFT calculations for the S22 benchmark set in ascending order. This table shows clearly that, independent of the dispersion corrections combined, the LC+vdW methods provide higher accuracies than those of semiempirical functionals. In particular, relatively poor results are produced by the vdW-DF methods, which combine pure functionals such as revPBE or the Hartree–Fock exchange integral with a vdW functional. Therefore, this table supports the idea that long-range exchange interactions play a crucial role in van der Waals bonds.

6.4 Relativistic Corrections

Relativistic effects are crucial for investigating chemistry. Although these effects have been neglected thus far, the electronic structures of molecules containing heavy atoms, i.e., the third row of the periodic table or lower, cannot be reproduced without them. In quantum chemistry, the relativistic effect indicates a *relativity theory*-based effect on electronic motions. Relativity theory is a collective term that includes both the special relativity (Einstein 1905) and general relativity (Einstein 1916) theories: The special relativity theory is based on *the principle of relativity establishing the invariance of physical laws for the Lorentz transformation* and *the principle of the constancy of the speed of light* for coordinate systems moving at a constant speed, in Minkovski's four-dimensional space-time coordinates. On the other hand, general relativity theory introduces *the equivalence principle, establishing the equivalence of gravity and inertia* to extend the special relativity theory to include accelerated systems. This theory is founded in Riemann space coordinates with a metric tensor, which is the solution of the gravity equation. Although electrons are accelerated under central forces, these are assumed to move uniformly. Therefore, the special relativity theory is usually more significant than the general theory in chemistry. The Lorentz transformation, on which the special relativity theory is based, was suggested by J. Larmor and H. A. Lorentz to solve the inconsistency of electrodynamics and classical mechanics. In this transformation, the motional states of systems are considered in space-time coordinates, which equate the variance of space coordinate (x, y, z) and time coordinate (ct): $(cdt)^2-(dx)^2-(dy)^2-(dz)^2 = 0$. Assuming that the speed of light is constant, independent of the coordinate system, Einstein proved that the distance between two points decreases in coordinate systems moving at high velocity. For example, the time-space coordinate of an inertia system S, (t, x, y, z), is correlated with the time-space coordinate of another inertia system S′ moving with relative velocity v along the x coordinate, (t', x', y', z'), as

$$(t', x', y', z') = \left(\frac{t - vx/c^2}{\sqrt{1 - v^2/c^2}}, \frac{x - vt}{\sqrt{1 - v^2/c^2}}, y, z \right). \tag{6.50}$$

What is important is that *the Schrödinger equation is relativistically incorrect.* Let us consider the time-dependent Schrödinger equation,

$$\hat{H}\Psi = \left[-\frac{1}{2} \left(\frac{\partial^2}{\partial x^2} + \frac{\partial^2}{\partial y^2} + \frac{\partial^2}{\partial z^2} \right) + V \right] \Psi = i \frac{\partial \Psi}{\partial t}. \tag{6.51}$$

In this equation, the kinetic energy on the left-hand side is the second derivative in terms of the space coordinate, while the right-hand side is the first derivative in terms of time. That is, the space and time coordinates are not equivalent to each other in this equation. This indicates that *the Schrödinger equation is not invariant for the Lorentz transformation, and therefore it is relativistically incorrect.*

The Dirac equation makes the kinetic energy part of the Schrödinger equation invariant for the Lorentz transformation (Dirac 1928):

$$\hat{H}^{\mathrm{D}}\Psi = \left[c\boldsymbol{\alpha} \cdot \hat{\mathbf{p}} + \boldsymbol{\beta} mc^2 + V \right] \Psi = i \frac{\partial \Psi}{\partial t}, \tag{6.52}$$

$$\beta = \begin{pmatrix} \mathbf{I} & \mathbf{0} \\ \mathbf{0} & -\mathbf{I} \end{pmatrix}, \ \mathbf{I} = \begin{pmatrix} 1 & 0 \\ 0 & 1 \end{pmatrix}, \tag{6.53}$$

$$\boldsymbol{\alpha}_w = \begin{pmatrix} \mathbf{0} & \boldsymbol{\sigma}_w \\ \boldsymbol{\sigma}_w & \mathbf{0} \end{pmatrix} \quad (w = x, y, z), \tag{6.54}$$

$$\boldsymbol{\sigma}_x = \begin{pmatrix} 0 & 1 \\ 1 & 0 \end{pmatrix}, \ \boldsymbol{\sigma}_y = \begin{pmatrix} 0 & -i \\ i & 0 \end{pmatrix}, \ \text{and } \boldsymbol{\sigma}_z = \begin{pmatrix} 1 & 0 \\ 0 & -1 \end{pmatrix}. \tag{6.55}$$

In order to understand the physical meaning of the equation, the electron mass m and the speed of light c, which are both unity in atomic units, are explicitly written in this and the next sections. The $\boldsymbol{\sigma}$ in Eq. (6.55) is called the *Pauli spin matrix* (Pauli 1925). This equation is invariant for the Lorentz transformation, because the momentum $\mathbf{p} = -i\nabla$ is the first derivative in terms of space. More importantly, this equation requires *four-component wavefunctions*,

$$\Psi = \begin{pmatrix} \Psi_\alpha^{\mathrm{L}} \\ \Psi_\beta^{\mathrm{L}} \\ \Psi_\alpha^{\mathrm{S}} \\ \Psi_\beta^{\mathrm{S}} \end{pmatrix}. \tag{6.56}$$

Note that the β term on the left-hand side of the Dirac equation, in Eq. (6.52) whose rest energy is 5.11×10^5 eV, is inconveniently large compared to chemical energies, which are on the order of several eV. Therefore, β is replaced with

$$\beta' = \begin{pmatrix} \mathbf{0} & \mathbf{0} \\ \mathbf{0} & -2\mathbf{I} \end{pmatrix}. \tag{6.57}$$

Even using this β', there must be continuum states below $-2mc^2$ in energy. Dirac interpreted that these continuum states are occupied by an infinite number of *positrons*. Although this interpretation, which is called *vacancy theory*, was subsequently disproven by quantum electrodynamics, the presence of positrons was confirmed in later experiments. Based on this, in Eq. (6.56), it is often interpreted that Ψ^{L}, *which is called the large-component wavefunction, indicates the wavefunction of electronic motions*, while Ψ^{S}, *which is called the small-component wavefunction, implies an electronic wavefunction affected by positronic motions*. Note, however, that both of these components are actually mixtures of these wavefunctions and that the degree of mixing increases as the systems become heavier.

It is interesting to compare the time-independent limit of the Dirac equation with the Schrödinger equation. The time-independent Dirac equation is given as

$$\left[c\boldsymbol{\alpha}\cdot\hat{\mathbf{p}}+\boldsymbol{\beta}'mc^2+V\right]\Psi=E\Psi \tag{6.58}$$

Since this equation is written as

$$c(\boldsymbol{\sigma}\cdot\hat{\mathbf{p}})\Psi^{\mathrm{S}}+V\Psi^{\mathrm{L}}=E\Psi^{\mathrm{L}} \tag{6.59}$$

$$c(\boldsymbol{\sigma}\cdot\hat{\mathbf{p}})\Psi^{\mathrm{L}}+(-2mc^2+V)\Psi^{\mathrm{S}}=E\Psi^{\mathrm{S}}, \tag{6.60}$$

the small-component wavefunction is represented by the large-component one, Ψ^{L}, using Eq. (6.60) such as

$$\Psi^{\mathrm{S}}=(E+2mc^2-V)^{-1}c(\boldsymbol{\sigma}\cdot\hat{\mathbf{p}})\Psi^{\mathrm{L}}=\hat{K}\cdot\frac{\boldsymbol{\sigma}\cdot\hat{\mathbf{p}}}{2mc}\Psi^{\mathrm{L}}, \tag{6.61}$$

where $\hat{K}$ is

$$\hat{K}=\left(1+\frac{E-V}{2mc^2}\right)^{-1}. \tag{6.62}$$

Based on Eq. (6.61), the time-independent Dirac equation is obtained as

$$\left[\frac{1}{2m}(\boldsymbol{\sigma}\cdot\hat{\mathbf{p}})\hat{K}(\boldsymbol{\sigma}\cdot\hat{\mathbf{p}})+V\right]\Psi^{\mathrm{L}}=E\Psi^{\mathrm{L}}. \tag{6.63}$$

Assuming that the speed of light would be infinity and using $\hat{K}=1$ and $(\boldsymbol{\sigma}\cdot\hat{\mathbf{p}})(\boldsymbol{\sigma}\cdot\hat{\mathbf{p}})=\hat{\mathbf{p}}^2$, this equation gives the (nonrelativistic) Schrödinger equation,

$$\left[\frac{\hat{\mathbf{p}}^2}{2m}+V\right]\Psi^{\mathrm{L}}=E\Psi^{\mathrm{L}}. \tag{6.64}$$

As clearly shown by comparing Eqs. (6.63) and (6.64), *the difference between the Dirac and Schrödinger equations appears only in the momentum*.

Even though the Dirac equation makes only the kinetic energy part of the Schrödinger equation relativistic, *the potential part of the latter equation is also relativistically incorrect*. The *electrostatic potential* (nuclear-electron potential

$1/r$ and electron–electron potential $1/r_{12}$) *depends only on space, and therefore it is not invariant for the Lorentz transformation.* This implies that the electrostatic force acts instantaneously, faster than the speed of light. However, long-range interactions should inherently act later in time compared to short-range interactions. *Taking into consideration this retardation effect of interactions requires a quite complicated QED formulation that includes photon exchange between charged particles.* The relatively simple formulation derived from the Taylor expansion of the electron–electron interaction potential in terms of the fine structure constant, $1/c$, is therefore usually used:

$$V_{\text{ee}}(\mathbf{r}_{12}) = \frac{1}{r_{12}} - \frac{1}{r_{12}}\left[\boldsymbol{\alpha}_1 \cdot \boldsymbol{\alpha}_2 + \frac{(\boldsymbol{\alpha}_1 \times \mathbf{r}_{12})(\boldsymbol{\alpha}_1 \times \mathbf{r}_{12})}{r_{12}^2}\right] \quad (6.65)$$

$$= \frac{1}{r_{12}} - \frac{1}{2r_{12}}\left[\boldsymbol{\alpha}_1 \cdot \boldsymbol{\alpha}_2 + \frac{(\boldsymbol{\alpha}_1 \cdot \mathbf{r}_{12})(\boldsymbol{\alpha}_1 \cdot \mathbf{r}_{12})}{r_{12}^2}\right], \quad (6.66)$$

where $\boldsymbol{\alpha}_i$ is the α matrix in Eq. (6.54) for the i-th particle. For the nuclear–electron potential, the relativistic correction is usually neglected, because it is on the order of $1/c^3$. The relativistic correction for the electron–electron interaction potential is called the *Breit interaction* (Breit 1929), in which $-\boldsymbol{\alpha}_1 \cdot \boldsymbol{\alpha}_2/r_{12}$ is separately called the Gaunt interaction (Gaunt 1929). This Breit interaction has the order of $1/c^2$, because it contains no $(1/c)$-order term. Even for this Breit interaction, the high-order perturbation terms are considerably complicated. Although these terms also require long computational times, they have a minimal effect on chemical reactions and properties. Therefore, these terms are usually neglected in quantum chemistry calculations, even for heavy atoms.

The relativistic correction for the kinetic energy in the Dirac equation is naturally applicable to the Kohn–Sham equation. This relativistic Kohn–Sham equation is called the *Dirac–Kohn–Sham equation* (Rajagopal 1978; MacDonald and Vosko 1979). The Dirac–Kohn–Sham equation is founded on the *Rajagopal–Callaway theorem*, which is the relativistic expansion of the Hohenberg–Kohn theorem on the basis of QED (Rajagopal and Callaway 1973). In this theorem, two theorems are contained: The first theorem proves that the four-component external potential, which is the vector-potential-extended external potential, is determined by the four-component current density, which is the current-density-extended electron density. On the other hand, the second theorem establishes the variational principle for every four-component current density. See Sect. 6.5 for vector potential and current density. Consequently, the solution of the Dirac–Kohn–Sham equation is represented by the four-component orbital. This four-component orbital is often called a "molecular spinor." However, this name includes no indication of "orbital," which is the solution of one-electron SCF equations; moreover, the targets of the calculations are not restricted to molecules. Therefore, in this book, this four-component orbital is called an *orbital spinor*. The Dirac–Kohn–Sham wavefunction is represented by the Slater determinant of orbital spinors (see Sect. 2.3). Following the Roothaan method (see Sect. 2.5), orbital spinors are represented by a linear combination of the four-component basis spinor functions, $\{\chi_p\}$,

$$\psi_i = \begin{pmatrix} \psi_i^{\mathrm{L}} \\ \psi_i^{\mathrm{S}} \end{pmatrix} = \begin{pmatrix} \sum_p C_{pi}^{\mathrm{L}} \chi_p^{\mathrm{L}} \\ \sum_p C_{pi}^{\mathrm{S}} \chi_p^{\mathrm{S}} \end{pmatrix}. \tag{6.67}$$

Similarly to the Kohn–Sham equation, in Eq. (4.13). the Dirac–Kohn–Sham equation is given using the above basis spinor function representation by

$$\mathbf{FC}_i = \epsilon_i \mathbf{SC}_i, \tag{6.68}$$

where $\mathbf{C}_i$ is the expansion coefficient of the basis spinor functions and ϵ_i is the i-th orbital spinor. In this equation, the Fock matrix, $\mathbf{F}$, and the overlap matrix, $\mathbf{S}$, have the following elements (Nakajima 2009):

$$F_{pq} = \begin{pmatrix} h_{pq}^{\mathrm{LL}} + J_{pq}^{\mathrm{LL}} + (V_{\mathrm{xc}})_{pq}^{\mathrm{LL}} & h_{pq}^{\mathrm{LS}} + (V_{\mathrm{xc}})_{pq}^{\mathrm{LS}}, \\ h_{pq}^{\mathrm{SL}} + (V_{\mathrm{xc}})_{pq}^{\mathrm{SL}} & h_{pq}^{\mathrm{SS}} + J_{pq}^{\mathrm{SS}} + (V_{\mathrm{xc}})_{pq}^{\mathrm{SS}} \end{pmatrix}, \tag{6.69}$$

and

$$S_{pq} = \begin{pmatrix} S_{pq}^{\mathrm{LL}} & 0 \\ 0 & S_{pq}^{\mathrm{SS}} \end{pmatrix}. \tag{6.70}$$

Assuming that X and X′ indicate the large-component (L) and small-component (S) wavefunctions, the terms in the matrix elements are given as follows: For one-electron terms,

$$h_{pq}^{\mathrm{LL}} = V_{pq}^{\mathrm{LL}}, \tag{6.71}$$

$$h_{pq}^{\mathrm{SL}} = c\Pi_{pq}^{\mathrm{SL}} = h_{pq}^{\mathrm{LS}*}, \tag{6.72}$$

and

$$h_{pq}^{\mathrm{SS}} = V_{pq}^{\mathrm{SS}} - 2c^2 S_{pq}^{\mathrm{SS}}, \tag{6.73}$$

where the elements of the overlap integral matrix, $\mathbf{S}^{\mathrm{XX}}$, the nuclear–electron potential integral matrix, $\mathbf{V}^{\mathrm{XX}}$, and the kinetic energy integral matrix, $\boldsymbol{\Pi}^{XX'}$, are

$$S_{pq}^{\mathrm{XX}} = \int d^3\mathbf{r} \chi_p^{\mathrm{X}*}(\mathbf{r}) \chi_q^{\mathrm{X}}(\mathbf{r}) \tag{6.74}$$

$$V_{pq}^{\mathrm{XX}} = \int d^3\mathbf{r} \chi_p^{\mathrm{X}*}(\mathbf{r}) V_{\mathrm{ne}} \chi_q^{\mathrm{X}}(\mathbf{r}) \tag{6.75}$$

and

$$\Pi_{pq}^{\mathrm{XX}'} = \int d^3\mathbf{r} \chi_p^{\mathrm{X}*}(\mathbf{r}) (\boldsymbol{\sigma} \cdot \hat{\mathbf{p}}) \chi_q^{\mathrm{X}'}(\mathbf{r}). \tag{6.76}$$

Remaining are the Coulomb interaction and exchange-correlation potential integrals:

$$J_{pq}^{\mathrm{XX}} = \sum_{r,s=1}^{n_{\mathrm{basis}}} \left(P_{pq}^{\mathrm{XX}} J_{pqrs}^{\mathrm{XX}} + P_{pq}^{\mathrm{X'X'}} J_{pqrs}^{\mathrm{XX'}} \right), \tag{6.77}$$

$$J_{pqrs}^{\mathrm{XX'}} = \int d^3\mathbf{r}_1 d^3\mathbf{r}_2 \chi_p^{\mathrm{X}*}(\mathbf{r}_1)\chi_q^{\mathrm{X'}*}(\mathbf{r}_2)\frac{1}{r_{12}}\chi_r^{\mathrm{X}}(\mathbf{r}_1)\chi_s^{\mathrm{X'}}(\mathbf{r}_2), \tag{6.78}$$

and

$$(V_{\mathrm{xc}})_{pq}^{\mathrm{XX'}} = \int d^3\mathbf{r}\chi_p^{\mathrm{X}*}(\mathbf{r})V_{\mathrm{xc}}\chi_q^{\mathrm{X'}}(\mathbf{r}), \tag{6.79}$$

and the density matrix:

$$P_{pq}^{\mathrm{XX'}} = \sum_i C_{pi}^{\mathrm{X}*} C_{qi}^{\mathrm{X'}}. \tag{6.80}$$

The Dirac–Kohn–Sham equation in Eq. (6.68) is solved using the above elements. However, the most stable electronic states are not obtained by directly solving this equation, because the variational principle for electronic states is not established due to the contribution of positronic states. To solve this problem, the large-component and small-component basis spinor functions are balanced using

$$\chi^{\mathrm{S}} = \frac{\boldsymbol{\sigma}\cdot\hat{\mathbf{p}}}{2c}\chi^{\mathrm{L}}, \tag{6.81}$$

which is called the *kinetic balance condition* (McLean and Lee 1982). This condition also contributes to the acceptance of the use of conventional basis functions for the basis spinor functions of the whole four-component wavefunction. However, the basis spinor functions of the small-component wavefunction requires double basis functions: the derivatives of the basis functions in addition to the basis functions themselves. The numbers of two-electron integrals for the large–small and small–small combinations are 8 and 16 times larger than that of the large–large combination. Consequently, *a Dirac–Kohn–Sham integral calculation requires a computational time that is approximately 25 times longer than that of the Kohn–Sham one*. Note that the relativistic effect is significant in heavy molecules containing many electrons. For this reason, calculations using the four-component Dirac–Kohn–Sham equation mostly target single-atom or several-atom systems, even at present.

The huge computational time of the Dirac–Kohn–Sham equation calculations is attributable to the use of the small-component wavefunction, producing the coupling of the large-component and small-component wavefunctions. *Two-component relativistic approximations* have been suggested to solve this problem and have now become mainstream. Breit applied the expansion of $\hat{K}$ for electronic motions sufficiently slower than the speed of light, as usual,

$$\hat{K} = \left(1 + \frac{E - V}{2mc^2}\right)^{-1} \approx 1 - \frac{E - V}{2mc^2} + \cdots \tag{6.82}$$

up to the order of $1/c^2$ to the Dirac equation, such as

$$\left[\frac{\hat{\mathbf{p}}^2}{2m} + V - \frac{\hat{\mathbf{p}}^4}{8m^3c^2} + \frac{Z\hat{\mathbf{s}} \cdot \hat{\mathbf{l}}}{2m^2c^2r^3} + \frac{Z\pi\delta(\mathbf{r})}{2m^2c^2}\right]\Psi^{\mathrm{L}} = E\Psi^{\mathrm{L}}, \tag{6.83}$$

where $\hat{\mathbf{l}}$ is the orbital angular momentum operator and δ is the delta function. Since this equation was first suggested by Pauli, it is called the *Breit–Pauli equation* (Breit 1929). This equation contains only the large-component wavefunction. On the left-hand side of Eq. (6.64), the additional terms compared to those of the Schrödinger equation, are, in sequence, the *mass velocity correction*, which is the effect of velocity on mass, the *spin–orbit interaction*, which is the magnetic interaction between electronic spin and the orbital, and the *Darwin correction*, which comes from the high-frequency vibrational electronic motion around the equilibrium geometry. The mass velocity and Darwin corrections are collectively called the *scalar relativistic correction*. However, the expansion of this $\hat{K}$ is often inappropriate due to the divergence of the potential ($V \to \infty$) near nuclei. To avoid this divergence, $\hat{K}$ is frequently replaced with $\hat{K}'$ defined by

$$\begin{aligned}(E + 2mc^2 - V)^{-1} &= (2mc^2 - V)^{-1}\left(1 + \frac{E}{2mc^2 - V}\right)^{-1} \\ &= (2mc^2 - V)^{-1}\hat{K}'.\end{aligned} \tag{6.84}$$

This enables us to avoid the divergence of the expansion due to $E/(2mc^2 - V) \ll 1$. Moreover, it is usually appropriate to approximate $\hat{K}'$ as unity. This approximation is called the *zeroth-order regular approximation (ZORA)*.

Another major two-component approximation is the *Foldy–Wouthuysen transformation* (Foldy and Wouthuysen 1950), which makes the large-component and small-component submatrices of the Dirac Hamiltonian matrix, $\hat{\mathbf{H}}_{\mathrm{D}}$, linear independent by a unitary transformation such as

$$\hat{\mathbf{H}}^{\mathrm{FW}} = U\hat{\mathbf{H}}^{\mathrm{D}}U^{\dagger} = \begin{pmatrix} \hat{H}^{\mathrm{L}} & 0 \\ 0 & \hat{H}^{\mathrm{S}} \end{pmatrix}. \tag{6.85}$$

Although $U = \exp(-imc^2)$ is used as the unitary operator in the original method, it does not work well in the use of potential V due to the presence of singular points. The application of the original transformation is limited to the case of free electrons ($V = 0$). For free electrons, this transformation is written as

$$U_0 = A_p(1 + \beta R_p), \tag{6.86}$$

$$A_p = \left(\frac{E_p + mc^2}{2E_p}\right)^{1/2}, \tag{6.87}$$

$$E_p = \left(\hat{\mathbf{p}}^2 c^2 + m^2 c^4\right)^{1/2}, \tag{6.88}$$

and

$$R_p = \frac{c\boldsymbol{\alpha} \cdot \hat{\mathbf{p}}}{E_p + mc^2}. \tag{6.89}$$

This transformation leads to the Hamiltonian operator for using a potential as

$$\hat{\mathbf{H}}^{\mathrm{FW}} = U_0 \hat{\mathbf{H}}^{\mathrm{D}} U_0^\dagger = \mathscr{E}_0 + \mathscr{E}_1 + \mathscr{O}_1, \tag{6.90}$$

$$\mathscr{E}_0 = \beta E_p - mc^2, \tag{6.91}$$

$$\mathscr{E}_1 = A_p(V + R_p V R_p)A_p, \tag{6.92}$$

and

$$\mathscr{O}_1 = \beta A_p \left[R_p, V\right] A_p. \tag{6.93}$$

This formulation includes the lowest-order off-diagonal term of the Hamiltonian matrix, $\mathscr{O}_1$, which leads to the production of singular points. However, these singular points can be eliminated by further appropriate unitary transformations. In this strategy, the *Douglas–Kroll transformation* solves the singular point problem of the Foldy–Wouthuysen transformation (Douglas and Kroll 1974):

$$\hat{\mathbf{H}}^{\mathrm{DK}} = U\hat{\mathbf{H}}^{\mathrm{D}}U^\dagger = \cdots U_4 U_3 U_2 U_1 \hat{\mathbf{H}}^{\mathrm{FW}} U_1^\dagger U_2^\dagger U_3^\dagger U_4^\dagger \cdots$$
$$= \sum_{i=0}^{\infty} \mathscr{E}_i. \tag{6.94}$$

Since it is actually impossible to sum the infinite series, this summation is terminated at particular numbers of unitary transformations: 2 in DK2, 3 in DK3 and so forth. This method is also called the Douglas–Kroll–Hess transformation, because it was revised by Hess and coworkers (Jansen and Hess 1989).

Finally, let us summarize the relativistic effects on the electronic states of atoms and molecules. 1s *orbitals contract*, because 1*s* orbital electrons moving at velocities close to that of light become much heavier. This causes *the contraction of higher s orbitals*, which are orthogonal to the 1*s* orbitals, and leads to more shielded nuclear charges. Moreover, the shielding of the nuclear charge leads to *the expansion in the sizes of the d, f, and higher-angular-momentum orbitals*, while the *p* orbitals do not become so large, due to spin–orbit interactions with *s* orbitals. This spin–orbit interaction transforms α-spin and β-spin orbitals to *orbital spinors* having mixed spin. *The shapes of the orbitals are also transformed* by the small-component wavefunction, coming from the interactions with positronic states. Furthermore, the finite velocity of electronic motions *changes the electron–electron interaction potentials*: e.g., it produces the Breit interaction.

6.5 Vector Potential Correction and Current Density

As shown in the previous section, several magnetic effects, such as the spin–orbit effect are included in the relativistic effect. To consider the effect of an external magnetic field, however, the *vector potential*, **A**, should be incorporated in the momentum operator (Jensen 2006),

$$\hat{\boldsymbol{\pi}} = \hat{\mathbf{p}} + \mathbf{A}, \tag{6.95}$$

where $\hat{\boldsymbol{\pi}}$ is called the *generalized momentum operator*. Since the vector potential is related to the external magnetic field, **B**, as

$$\mathbf{B} = \nabla \times \mathbf{A}, \tag{6.96}$$

it is represented as

$$\mathbf{A} = \frac{1}{2}\mathbf{B} \times (\mathbf{r} - \mathbf{R}^{\mathrm{G}}), \tag{6.97}$$

where $\mathbf{R}^{\mathrm{G}}$ is the center of the vector potential, called the *gauge center*, which is usually set at the center of mass. Note that although the vector potential does not depend on the position of the gauge center in principle, it always does so in approximate calculations. The time-independent Dirac equation using the generalized momentum operator is given as

$$\left[\frac{1}{2m}(\boldsymbol{\sigma} \cdot \hat{\boldsymbol{\pi}})\hat{K}(\boldsymbol{\sigma} \cdot \hat{\boldsymbol{\pi}}) + V\right]\Psi^{\mathrm{L}} = E\Psi^{\mathrm{L}}. \tag{6.98}$$

At the nonrelativistic limit ($c \to 0$), this equation becomes

$$\left\{\frac{1}{2m}\left[\hat{\boldsymbol{\pi}}^2 + i\boldsymbol{\sigma} \cdot (\boldsymbol{\pi} \times \hat{\boldsymbol{\pi}})\right] + V\right\}\Psi^{\mathrm{L}} = E\Psi^{\mathrm{L}}. \tag{6.99}$$

Since the term in parentheses in the imaginary term on the left-hand side is derived as

$$\begin{aligned}(\hat{\boldsymbol{\pi}} \times \hat{\boldsymbol{\pi}})\Psi^{\mathrm{L}} &= (\hat{\mathbf{p}} \times \mathbf{A} + \mathbf{A} \times \hat{\mathbf{p}})\,\Psi^{\mathrm{L}} \\ &= -i\nabla \times (\mathbf{A}\Psi^{\mathrm{L}}) - i\mathbf{A} \times (\nabla\Psi^{\mathrm{L}}) = -i\mathbf{B}\Psi^{\mathrm{L}},\end{aligned} \tag{6.100}$$

the Dirac equation at the nonrelativistic limit is given by

$$\left(\frac{\hat{\boldsymbol{\pi}}^2}{2m} + V + \frac{\boldsymbol{\sigma}\cdot\mathbf{B}}{2m}\right)\Psi^{\mathrm{L}} = E\Psi^{\mathrm{L}}. \tag{6.101}$$

The extended magnetic term on the left-hand side, which is called the *Zeeman interaction term*, leads to the Zeeman effect, which is a splitting of spectral lines under the influence of a magnetic field. Meanwhile, it is easily proven for non-hybridized electron spins in the nonrelativistic case that the Pauli spin matrix, $\boldsymbol{\sigma}$, in Eq. (6.55) is just twice the spin operator, $\hat{\mathbf{s}}$, in Eq. (2.91). The Zeeman interaction term is, therefore, written as

$$\frac{\boldsymbol{\sigma}\cdot\mathbf{B}}{2m} = g_{\mathrm{e}}\mu_{\mathrm{B}}\hat{\mathbf{s}}\cdot\mathbf{B}, \tag{6.102}$$

where $\mu_{\mathrm{B}} = 1/2m$ is called the *Bohr magneton*. In Eq. (6.102), g_{e}, called the *Landé g factor* is simply 2, although it is actually found to shift slightly to 2.0023 due to the quantum field fluctuation effect in quantum electrodynamics. This term causes the spin splitting observed in the spectra of nuclear magnetic resonance (NMR) and electron spin resonance (ESR).

Next, let us consider the kinetic energy in the first term on the left-hand side of Eq. (6.101). The square of the generalized momentum operator, $\hat{\boldsymbol{\pi}}$, is derived as

$$\hat{\boldsymbol{\pi}}^2 = \hat{\mathbf{p}}^2 + \hat{\mathbf{p}}\cdot\mathbf{A} + \mathbf{A}\cdot\hat{\mathbf{p}} + \mathbf{A}^2, \tag{6.103}$$

$$(\hat{\mathbf{p}}\cdot\mathbf{A})\,\Psi^{\mathrm{L}} = -i(\nabla\cdot\mathbf{A})\Psi^{\mathrm{L}} = -i\left[\mathbf{A}\cdot(\nabla\Psi^{\mathrm{L}}) + \Psi^{\mathrm{L}}(\nabla\cdot\mathbf{A})\right], \tag{6.104}$$

$$\mathbf{A}\cdot\hat{\mathbf{p}} = \left[\frac{1}{2}\mathbf{B}\times(\mathbf{r}-\mathbf{R}^{\mathrm{G}})\right]\cdot\hat{\mathbf{p}} = \frac{1}{2}\mathbf{B}\cdot(\mathbf{r}-\mathbf{R}^{\mathrm{G}})\times\hat{\mathbf{p}}, \tag{6.105}$$

and

$$\begin{aligned}\mathbf{A}^2 &= \left[\frac{1}{2}\mathbf{B}\times(\mathbf{r}-\mathbf{R}^{\mathrm{G}})\right]^2 \\ &= \frac{1}{4}\left\{\mathbf{B}^2\cdot(\mathbf{r}-\mathbf{R}^{\mathrm{G}})^2 - \left[\mathbf{B}\cdot(\mathbf{r}-\mathbf{R}^{\mathrm{G}})\right]^2\right\}.\end{aligned} \tag{6.106}$$

Fortunately, the Zeeman interaction term in Eq. (6.102) can be assembled with the inner product of the vector potential and the momentum in Eq. (6.105) by using the *magnetization density operator*, $\hat{\mathbf{m}}$, corresponding to the *magnetic dipole moment*, $\mathbf{m}$, whose expectation value is the reverse sign of the first energy derivative in terms of magnetic field, as

$$\left[g_{\mathrm{e}}\mu_{\mathrm{B}}\hat{\mathbf{s}} + \frac{1}{4m}(\mathbf{r}-\mathbf{R}^{\mathrm{G}})\times\hat{\mathbf{p}}\right]\cdot\mathbf{B} = -\int d^3\mathbf{r}\,\hat{\mathbf{m}}\cdot\mathbf{B}. \tag{6.107}$$

For usual nonrelativistic wavefunctions represented by the Slater determinant, the right-hand side of Eq. (6.104) is derived using the *paramagnetic current density*,

$$\mathbf{j}_{\mathrm{p}}(\mathbf{r}) = \frac{i}{2m} \sum_{n} \left[\phi_n^*(\mathbf{r}) \nabla \phi_n(\mathbf{r}) - \phi_n(\mathbf{r}) \nabla \phi_n^*(\mathbf{r}) \right], \tag{6.108}$$

as

$$\frac{(\hat{\mathbf{p}} \cdot \mathbf{A})}{2m} \Psi^{\mathrm{L}} = -\left[\int d^3\mathbf{r}\, \mathbf{j}_{\mathrm{p}}(\mathbf{r}) \cdot \mathbf{A}(\mathbf{r}) \right] \Psi^{\mathrm{L}} \tag{6.109}$$

(see, e.g., Rajagopal and Callaway 1973). This paramagnetic current density, $\mathbf{j}_{\mathrm{p}}$, is not the total current density, $\mathbf{j}$. Actually, the total current density also contains the *diamagnetic current density*, $\mathbf{j}_{\mathrm{d}}$, and *magnetization current density*, $\mathbf{j}_{\mathrm{m}}$, such as

$$\begin{aligned} \mathbf{j}(\mathbf{r}) &= \mathbf{j}_{\mathrm{p}}(\mathbf{r}) + \mathbf{j}_{\mathrm{d}}(\mathbf{r}) + \mathbf{j}_{\mathrm{m}}(\mathbf{r}) \\ &= \mathbf{j}_{\mathrm{p}}(\mathbf{r}) - \frac{1}{m} \rho(\mathbf{r}) \mathbf{A}(\mathbf{r}) + \nabla \times \mathbf{m}(\mathbf{r}). \end{aligned} \tag{6.110}$$

In summary, the Dirac equation under the influence of an external magnetic field is written at the nonrelativistic limit as

$$\left[\frac{\hat{\mathbf{p}}^2 + \mathbf{A}^2}{2m} + V - \int d^3\mathbf{r} \left(\mathbf{j}_{\mathrm{p}} \cdot \mathbf{A} + \hat{\mathbf{m}} \cdot \mathbf{B} \right) \right] \Psi^{\mathrm{L}} = E \Psi^{\mathrm{L}}. \tag{6.111}$$

Vignale and Rasolt derived *the Dirac–Kohn–Sham equation incorporating the vector potential* from the nonrelativistic Dirac equation neglecting the magnetic effect, i.e., the Zeeman interaction term, in Eq. (6.101) as (Vignale and Rasolt 1987, 1988)

$$\left[\frac{(\hat{\mathbf{p}} + \mathbf{A} + \mathbf{A}_{\mathrm{xc}})^2}{2m} + V_{\mathrm{ext}} + \sum_{j}^{n} \hat{J}_j + V_{\mathrm{xc}} \right] \phi_i = \epsilon_i \phi_i. \tag{6.112}$$

In this equation, the exchange-correlation energy functional, E_{xc}, is assumed to be a functional of the electron density and paramagnetic current density, and the corresponding exchange-correlation potential, V_{xc}, and vector potential, $\mathbf{A}_{\mathrm{xc}}$, are represented as

$$V_{\mathrm{xc}} = \left. \frac{\delta E_{\mathrm{xc}}[\rho, \mathbf{j}_{\mathrm{p}}]}{\delta \rho(\mathbf{r})} \right|_{\delta \mathbf{j}_{\mathrm{p}} = 0}, \tag{6.113}$$

and

$$\mathbf{A}_{\mathrm{xc}} = \left. \frac{\delta E_{\mathrm{xc}}[\rho, \mathbf{j}_{\mathrm{p}}]}{\delta \mathbf{j}_{\mathrm{p}}(\mathbf{r})} \right|_{\delta \rho = 0}, \tag{6.114}$$

respectively. The theory based on Eq. (6.112) is called (strictly) the *current density functional theory*. Note that *this equation is not invariant for the gauge transformation of the vector potential* (Takada 2009). In quantum mechanics, the gauge transformation indicates the transformation of potential, which is commutative with the Hamiltonian operator. According to quantum electrodynamics, the transformation of the vector potential,

$$\mathbf{A} \rightarrow \mathbf{A} + \nabla \Lambda \tag{6.115}$$

is a gauge transformation. The arbitrary scalar function, Λ, and its derivative, $\nabla\Lambda$, are called *gauge*. However, for the gauge transformation in Eq. (6.115), the paramagnetic current density, $\mathbf{j}_{\mathrm{p}}$, is not invariant, because it creates a diamagnetic current density in Eq. (6.110) such as

$$\mathbf{j}_{\mathrm{p}}(\mathbf{r}) \rightarrow \mathbf{j}_{\mathrm{p}}(\mathbf{r}) + \frac{1}{m}\rho(\mathbf{r})\nabla\Lambda(\mathbf{r}). \tag{6.116}$$

Exchange-correlation functionals using current density are, therefore, also not invariant for the gauge transformation. To ensure gauge invariance, the gauge-invariant *vorticity*,

$$\boldsymbol{v}(\mathbf{r}) = -\nabla \times \left(\frac{\mathbf{j}_{\mathrm{p}}(\mathbf{r})}{\rho(\mathbf{r})}\right) \rightarrow -\nabla \times \left(\frac{\mathbf{j}_{\mathrm{p}}(\mathbf{r})}{\rho(\mathbf{r})} + \frac{1}{m}\nabla\Lambda(\mathbf{r})\right) = \boldsymbol{v}(\mathbf{r}), \tag{6.117}$$

is often used in exchange-correlation functionals, instead of the paramagnetic current density. Using the vorticity, the exchange-correlation potential and vector potential are given as

$$V_{\mathrm{xc}} = \left.\frac{\delta E_{\mathrm{xc}}[\rho, \boldsymbol{v}]}{\delta\rho(\mathbf{r})}\right|_{\delta\boldsymbol{v}=0} - \mathbf{A}_{\mathrm{xc}} \cdot \frac{\mathbf{j}_{\mathrm{p}}(\mathbf{r})}{\rho(\mathbf{r})}, \tag{6.118}$$

and

$$\mathbf{A}_{\mathrm{xc}} = \left.\frac{\nabla}{\rho(\mathbf{r})} \times \frac{\delta E_{\mathrm{xc}}[\rho, \boldsymbol{v}]}{\delta\boldsymbol{v}(\mathbf{r})}\right|_{\delta\rho=0}. \tag{6.119}$$

So far, the current density functional has attracted attention, not in the context of the response to a magnetic field, as mentioned above, *but to an electric field.* The time-dependent Kohn–Sham equation in Eq. (4.27) incorporating the time-dependent vector potential, $\mathbf{A}_{\mathrm{eff}}$, is written as

$$i\frac{\partial}{\partial t}\phi_i(\mathbf{r}, t) = \left[-\frac{1}{2}(-i\nabla + \mathbf{A}_{\mathrm{eff}}(\mathbf{r}, t))^2 + V_{\mathrm{eff}}[\mathbf{r}, t; \rho(\mathbf{r}, t)]\right]\phi_i(\mathbf{r}, t). \tag{6.120}$$

Vignale and Kohn proved that this time-dependent vector potential is Fourier-transformed ($t \rightarrow \omega$) using the current density $\mathbf{j}$, which is usually the paramagnetic current density $\mathbf{j}_{\mathrm{p}}$, to

$$\mathbf{A}_{\mathrm{eff}} = \mathbf{A}_{\mathrm{ext}} + \mathbf{A}_{\mathrm{xc}}, \tag{6.121}$$

$$\mathbf{A}_{\mathrm{xc}}(\mathbf{r},\omega) = \int d^3\mathbf{r}' f_{\mathrm{xc}}(\mathbf{r},\mathbf{r}',\omega)\cdot\mathbf{j}(\mathbf{r}',\omega), \tag{6.122}$$

where f_{xc} is the exchange-correlation integral kernel in Eq. (4.31), and suggested that *the problems in the response property calculations* (see Sect. 6.1), *which are presumed to come from the locality of functionals, might be solved by using this time-dependent vector potential* (Vignale and Kohn 1996). Even considering the time-dependent vector potential, the time-dependent Kohn–Sham equation has the same form as Eq. (4.36),

$$\boldsymbol{\Omega}\mathbf{F}_{ia\sigma} = \omega_{ia\sigma}^2 \mathbf{F}_{ia\sigma}, \tag{6.123}$$

$$\begin{aligned}\Omega_{ia\sigma,jb\tau}^{\mathrm{singlet}} &= \delta_{\sigma\tau}\delta_{ij}\delta_{ab}\left(\epsilon_{a\sigma}-\epsilon_{i\sigma}\right)^2 \\ &\quad +2\left(\epsilon_{a\sigma}-\epsilon_{i\sigma}\right)^{1/2}\left(K_{ia,jb}^{\sigma\sigma}+K_{ia,jb}^{\sigma\sigma'}\right)\left(\epsilon_{b\sigma}-\epsilon_{j\sigma}\right)^{1/2},\end{aligned} \tag{6.124}$$

and

$$\begin{aligned}\Omega_{ia\sigma,jb\tau}^{\mathrm{triplet}} &= \delta_{\sigma\tau}\delta_{ij}\delta_{ab}\left(\epsilon_{a\sigma}-\epsilon_{i\sigma}\right)^2 \\ &\quad +2\left(\epsilon_{a\sigma}-\epsilon_{i\sigma}\right)^{1/2}\left(K_{ia,jb}^{\sigma\sigma}-K_{ia,jb}^{\sigma\sigma'}\right)\left(\epsilon_{b\sigma}-\epsilon_{j\sigma}\right)^{1/2}.\end{aligned} \tag{6.125}$$

However, the response matrix, $\mathbf{F}_{ia\sigma}$, in Eq. (4.39) is replaced with

$$F_{ia\sigma} = \left(\epsilon_{a\sigma}-\epsilon_{i\sigma}\right)^{-1/2}\left(X_{ia\sigma}-X_{ai\sigma}\right), \tag{6.126}$$

$$\begin{aligned}X_{ia\sigma}(\omega) &= \frac{-1}{\omega+\left(\epsilon_{a\sigma}-\epsilon_{i\sigma}\right)}\left[\int d^3\mathbf{r}\phi_{i\sigma}^*(\mathbf{r})\delta\left(\sum_i^n \hat{J}_i + V_{\mathrm{xc}}\right)(\mathbf{r},\omega)\phi_{a\sigma}(\mathbf{r})\right. \\ &\quad \left. +\frac{\omega}{\epsilon_{a\sigma}-\epsilon_{i\sigma}}\int d^3\mathbf{r}\phi_{i\sigma}^*(\mathbf{r})\hat{\mathbf{j}}\phi_{a\sigma}(\mathbf{r})\delta\mathbf{A}_{\mathrm{eff},\sigma}(\mathbf{r},\omega)\right],\end{aligned} \tag{6.127}$$

and $K_{ia,jb}^{\sigma\tau}$ is substituted by

$$\begin{aligned}K_{ia,jb}^{\sigma\tau} &= \langle ib|aj\rangle^{\sigma\tau} \\ &\quad +\iint d^3\mathbf{r}_1 d^3\mathbf{r}_2 \phi_{i\sigma}^*(\mathbf{r}_1)\phi_{b\sigma}^*(\mathbf{r}_2) f_{\mathrm{xc}}^{\sigma\tau}(\mathbf{r}_1,\mathbf{r}_2)\phi_{a\tau}(\mathbf{r}_1)\phi_{j\tau}(\mathbf{r}_2) \\ &\quad +\left(\frac{\omega}{\epsilon_{a\sigma}-\epsilon_{i\sigma}}\right)^2\iint d^3\mathbf{r}_1 d^3\mathbf{r}_2 \phi_{i\sigma}^*(\mathbf{r}_1)\hat{\mathbf{j}}\phi_{a\sigma}(\mathbf{r}_1) \\ &\quad \times f_{\mathrm{xc}}^{\sigma\tau}(\mathbf{r},\mathbf{r}',\omega)\phi_{j\tau}(\mathbf{r}_2)\hat{\mathbf{j}}\phi_{b\tau}^*(\mathbf{r}_2).\end{aligned} \tag{6.128}$$

By the calculations of *the time-dependent current density functional theory* using this equation, accurate excitation energies are obtained for some $\pi \rightarrow \pi^*$ excitations (van Faassen and de Boeij 2004). Meanwhile, however, it has been found that quite poor excitation energies are produced for the excitations of some types of molecules. On the other hand, calculations of the adiabatic excitation energy benchmark set, containing 109 molecules, show that the vector potential correction hardly affects the calculated excitation energies (Bates and Furche 2012).

The vector potential is also incorporated in the coupled-perturbed Kohn–Sham method. In this method, the following term is supplemented to matrix $(\mathbf{F}')$ in Eq. (4.61):

$$(\mathbf{F}')_{ia}^{\text{vector}} = \frac{1}{2}\int d^3\mathbf{r}\phi_a^*(\mathbf{r})\left(\hat{\mathbf{j}}\cdot\mathbf{A}_{\text{xc}}^{\text{viscoel}}(\mathbf{r},\omega)+\mathbf{A}_{\text{xc}}^{\text{viscoel}}(\mathbf{r},\omega)\cdot\hat{\mathbf{j}}\right)\phi_i(\mathbf{r}), \tag{6.129}$$

where the current density operator, $\hat{\mathbf{j}}$, is given as

$$\hat{\mathbf{j}} = -\frac{i}{2}\left(\nabla - \nabla^{\dagger}\right), \tag{6.130}$$

and $\mathbf{A}_{\text{xc}}^{\text{viscoel}}$ is called the exchange-correlation vector potential in a viscoelastic field. For a detailed formulation of the vector potential incorporating the viscoelasticity of the electron liquid, see, e.g., van Faassen et al. (2003). By applying this *coupled perturbed current density functional theory* to the polarizability calculations of long-chain molecules, it has been found that the overestimation of these polarizabilities is significantly improved. However, other optical response properties have so far never been calculated, differently from the long-range corrected Kohn–Sham calculations (see Sect. 6.1).

As an interesting application of the current density correction, let us finally examine the orbital energies of atoms. *A Kohn–Sham calculation incorrectly produces different orbital energies for degenerate atomic orbitals, depending on the magnetic quantum numbers.* Becke suggested that this degeneracy breaking is improved by incorporating the current density in functionals (Becke 2002). In Table 6.2, the calculated orbital energy differences in the outermost p orbitals, which should be degenerate, are displayed for each atom (Becke 2002; Maximoff et al. 2004). The functionals with a "j"-prefix are current-density-corrected, having been explicitly corrected by a $\mathbf{j}^2/\rho$ term. In this table, the current-density-corrected functionals give much smaller orbital energy differences than those of the uncorrected ones. This result has attracted attention, because it had so far been unclear why these orbitals are not degenerate. However, the transition energies between atomic orbitals were calculated by the exact time-dependent current density functional theory thereafter, as shown in Table 6.3. This table indicates that the current density terms in the exact theory have the effect of overestimating the $2s \rightarrow 2p$ orbital transition energies. Therefore, *even though the current density certainly affects the degenerate p orbital energies of atoms, it is open to question whether the lack of current density*

Table 6.2 Atomic orbital energy differences in the outermost p orbitals of magnetic quantum numbers $m_l = 1$ and 0 ($E(m_l = 1) - E(m_l = 0)$) in kcal/mol

Atom	LDAXC	B88XC	PBEXC	jBRX	jPBEXC
B	1.0	2.7	2.9	0.6	0.1
C	0.3	2.5	2.7	0.4	−0.2
O	1.6	4.6	6.1	0.9	−0.7
F	0.6	4.1	5.5	0.7	−0.7
Al	0.4	1.1	1.7	0.2	0.3
Si	−0.2	0.6	1.3	0.0	−0.1
S	0.2	1.3	2.8	0.1	0.2
Cl	−0.5	0.8	2.2	0.0	−0.2

The "jBRX" indicates the use of the Becke–Roussel (BR) functional (Becke and Roussel 1989) containing a current density term (j). Excerpt from Becke (2002)

Table 6.3 Calculated transition energies between the orbitals of the Be atom in eV

Transition	Expt.	ALDA1	jALDA1	ALDA2	jALDA2
$2s \rightarrow 2p$	5.27	5.07	6.24	4.86	5.62
$2s \rightarrow 3s$	6.77	5.62	5.67	5.65	5.63

The "jALDA1" indicates the first-version adiabatic LDA (ALDA) functional corrected by incorporating vector potentials (j). The ET-pVQZ basis functions are used. Excerpt from van Faassen and de Boeij (2004)

effects causes the orbital degeneracy breaking in the usual Kohn–Sham method. In addition, the breaking of gauge invariance mentioned above might also affect the calculated transition energies, because none of the methods mentioned above uses the vorticity in Eq. (6.117).

References

Andersson, Y., Langreth, D.C., Lundqvist, B.I.: Phys. Rev. Lett. **76**, 102–105 (1996)
Antony, J., Grimme, S.: Phys. Chem. Chem. Phys. **8**, 5287–5293 (2006)
Arai, M., Fujiwara, T.: Phys. Rev. B **51**, 1477–1489 (1995)
Bates, J.E., Furche, F.: J. Chem. Phys. **137**, 164105(1–10) (2012)
Becke, A.D.: J. Chem. Phys. **117**, 6935–6938 (2002)
Becke, A.D., Johnson, E.R.: J. Chem. Phys. **123**, 154101(1–9) (2005a)
Becke, A.D., Johnson, E.R.: J. Chem. Phys. **125**, 154105(1–5) (2005b)
Becke, A.D., Roussel, M.R.: Phys. Rev. A **39**, 3761–3767 (1989)
Bleiziffer, P., Heßelmann, A., Görling, A.: J. Chem. Phys. **136**, 134102(1–15) (2012)
Breit, G.: Phys. Rev. **34**, 553–573 (1929)
Casimir, H., Polder, D.: Phys. Rev. **73**, 360–372 (1948)
Chai, J.-D., Head-Gordon, M.: J. Chem. Phys. **128**, 084106(1–15) (2008a)
Chai, J.-D., Head-Gordon, M.: Phys. Chem. Chem. Phys. **10**, 6615–6620 (2008b)
Chai, J.-D., Head-Gordon, M.: J. Chem. Phys. **131**, 174105(1–13) (2009)
Champagne, B., Perpéte, E.A., Jacquemin, D.: J. Phys. Chem. A **104**, 4755–4763 (2000)
Cohen, A.J., Mori-Sanchez, P., Yang, W.: J. Chem. Phys. **126**, 191109(1–5) (2007)

Dion, M., Rydberg, H., Schröder, E., Langreth, D.C., Lundqvist, B.I.: Phys. Rev. Lett. **92**, 246401(1–4) (2004)
Dirac, P.A.M.: Proc. Roy. Soc. A **117**, 610–624 (1928)
Dirac, P.A.M.: Camb. Phil. Soc. **26**, 376–385 (1930)
Dobson, J.F., Dinte, B.P.: Phys. Rev. Lett. **76**, 1780–1783 (1996)
Douglas, M., Kroll, N.M.: Ann. Phys. **82**, 89–155 (1974)
Dreizler, R.M., Gross, E.K.U.: Density-Functional Theory An Approach to the Quantum Many-Body Problem. Springer, Berlin (1990)
Dreuw, A., Head-Gordon, M.: J. Am. Chem. Soc. **126**, 4007–4016 (2004)
Dreuw, A., Weisman, J.L., Head-Gordon, M.: J. Chem. Phys. **119**, 2943–2946 (2003)
Einstein, A.: Ann. Phys. **17**, 891–921 (1905)
Einstein, A.: Ann. Phys. **49**, 769–822 (1916)
Ernzerhof, M., Perdew, J.P.: J. Chem. Phys. **109**, 3313–3320 (1998)
Fermi, E., Amaldi, E.: Accad. Ital. Rome, **6**, 117–149 (1934)
Foldy, L.L., Wouthuysen, S.A.: Phys. Rev. **78**, 29–36 (1950)
Gaunt, I.P.: Proc. R. Soc. Lond. A **124**, 163–176 (1929)
Gerber, I.G., Angyan, J.G., Marsman, M., Kresse, G.: J. Chem. Phys. **125**, 054101(1–6) (2007)
Giese, T.J., Audette, V.M., York, D.M.: J. Chem. Phys. **119**, 2618–2622 (2003)
Gräfenstein, J., Kraka, E., Cremer, D.: Phys. Chem. Chem. Phys. **6**, 1096–1112 (2004)
Grimme, S.: J. Chem. Phys. **124**, 034108–034116 (2006)
Grimme, S., Antony, J., Ehrlich, S., Krieg, H.: J. Chem. Phys. **132**, 154104 (2010)
Hartree, D.R.: Math. Proc. Camb. Phil. Soc. **24**, 89–132, 426–437 (1928)
Iikura, H., Tsuneda, T., Yanai, T., Hirao, K.: J. Chem. Phys. **115**(8), 3540–3544 (2001)
Israelachvili, J.N.: Intermolecular and Surface Forces. Academic, London (1992)
Jansen, G., Hess, B.A.: Phys. Rev. A **39**, 6016–6017 (1989)
Jensen, F.: Introduction to Computational Chemistry. Wiley, Chichester (2006)
Johnson, B.G., Gonzales, C.A., Gill, P.M.W., Pople, J.A.: Chem. Phys. Lett. **221**, 100–108 (1994)
Jurecka, P., Sponer, J., Cerny, J., Hobza, P.: Phys. Chem. Chem. Phys. **8**, 1985–1993 (2006)
Kamiya, M., Tsuneda, T., Hirao, K.: J. Chem. Phys. **117**, 6010–6015 (2002)
Kamiya, M., Sekino, H., Tsuneda, T., Hirao, K.: J. Chem. Phys. **122**, 234111(1–10) (2005)
Langreth, D.C., Perdew, J.P.: Solid State Commun. **17**, 1425–1429 (1975)
Lennard-Jones, J.E.: Proc. R. Soc. Lond. A **106**, 463–477 (1924)
Livshits, E., Baer, R.: Phys. Chem. Chem. Phys. **9**, 2932–2941 (2007)
London, F.W.: Z. Phys. **63**, 245–279 (1930)
MacDonald, A.H., Vosko, S.H.: J. Phys. C: Solid State Phys. **12**, 2977–2990 (1979)
Maximoff, S.N., Ernzerhof, M., Scuseria, G.E.: J. Chem. Phys. **120**, 2105–2109 (2004)
McLean, A.D., Lee, Y.S.: In: Carbó, R. (ed.) Current Aspects of Quantum Chemistry 1981. Elsevier, Amsterdam (1982)
McWeeny, R.: Methods of Molecular Quantum Mechanics, 2nd edn. Academic, San Diego (1992)
Nakajima, T.: Quantum Chemistry - An Introduction to Molecular Orbital Theory (Japanese). Shokabo, Tokyo (2009)
Nakata, A., Tsuneda, T.: J. Chem. Phys. **139**, 064102(1–10) (2013)
Nakata, A., Tsuneda, T., Hirao, K.: J. Phys. Chem. A **114**, 8521–8528 (2010)
Orszag, S.A.: Stud. Appl. Math. **51**, 253–259 (1972)
Pauli, W.: Z. Phys. **31**, 765–783 (1925)
Perdew, J.P., Zunger, A.: Phys. Rev. B **23**, 5048–5079 (1981)
Pernal, K., Podeszwa, R., Patkowski, K., Szalewicz, K.: Phys. Rev. Lett. **103**, 263201(1–4) (2009)
Rajagopal, A.K.: J. Phys. C: Solid State Phys. **11**, L943–L948 (1978)
Rajagopal, A.K., Callaway, J.: Phys. Rev. B **7**, 1912–1919 (1973)
Riley, K.E., Pitonak, M., Jurecka, P., Hobza, P.: Chem. Rev. **110**, 5023–5063 (2010)
Sato, T., Nakai, H.: J. Chem. Phys. **131**, 224104(1–12) (2009)
Sato, T., Tsuneda, T., Hirao, K.: J. Chem. Phys. **126**, 234114(1–12) (2007)
Savin, A.: In: Seminario, J.J. (ed.) Recent Developments and Applications of Modern Density Functional Theory. Elsevier, Amsterdam (1996)

Schwabe, T., Grimme, S.: Phys. Chem. Chem. Phys. **9**, 3397–3406 (2007)
Song, J.-W., Hirosawa, T., Tsuneda, T., Hirao, K.: J. Chem. Phys. **126**, 154105(1–7) (2007a)
Song, J.-W., Tokura, S., Sato, T., Watson, M.A., Hirao, K.: J. Chem. Phys. **127**, 154109(1–6) (2007b)
Starkschall, G., Gordon, R.: J. Chem. Phys. **56**, 2801–2806 (1972)
Svane, A., Gunnarsson, O.: Phys. Rev. Lett. **65**, 1148–1151 (1990)
Takada, Y.: Many-body Problems, Asakura Physics Series No. 15 (Japanese). Asakura, Tokyo (2009)
Tawada, Y., Tsuneda, T., Yanagisawa, S., Yanai, T., Hirao, K.: J. Chem. Phys. **120**, 8425–8433 (2004)
Toulouse, J., Zhu, W., Savin, A., Jansen, G., Angyan, J.G.: J. Chem. Phys. **135**, 084119(1–8) (2011)
Tozer, D.J., Handy, N.C.: J. Chem. Phys. **109**, 10180–10189 (1998)
Tsuneda, T., Kamiya, M., Hirao, K.: J. Comput. Chem. **24**, 1592–1598 (2003)
Tsuneda, T., Kamiya, M., Morinaga, N., Hirao, K.: J. Chem. Phys. **114**, 6505–6513 (2001)
Tsuneda, T., Sato, T.: BUTSURI **64**, 291–296 (2009)
van Faassen, M., de Boeij, P.L.: J. Chem. Phys. **120**, 8353–8363 (2004)
van Faassen, M., de Boeij, P.L., van Leeuwen, R., Berger, J.A., Snijders, J.G.: J. Chem. Phys. **118**, 1044–1053 (2003)
van Gisbergen, S.J.A., Kootstra, F., Schipper, P.R.T., Gritsenko, O.V., Snijders, J.G., Baerends, E.J.: Phys. Rev. A **57**, 2556–2571 (1998)
Vignale, G., Kohn, W.: Phys. Rev. Lett. **77**, 2037–2040 (1996)
Vignale, G., Rasolt, M.: Phys. Rev. Lett. **59**, 2360–2363 (1987)
Vignale, G., Rasolt, M.: Phys. Rev. B **37**, 10685–10696 (1988)
Vydrov, O.A., van Voorhis, T.: J. Chem. Phys. **130**, 104105(1–7) (2009)
Vydrov, O.A., Van Voorhis, T.: J. Chem. Phys. **132**, 164113(1–10) (2010)
Vydrov, O.A., Heyd, J., Krukau, A., Scuseria, G.E.: J. Chem. Phys. **125**, 074106(1–9) (2006)
Williams, H.L., Chabalowski, C.F.: J. Phys. Chem. A **105**, 646–659 (2001)
Yanai, T., Tew, D.P., Handy, N.C.: Chem. Phys. Lett. **91**, 51–57 (2004)
Zhao, Y., Truhlar, D.G.: Theor. Chem. Acc. **120**, 215–241 (2008)
Zhu, W., Toulouse, J., Savin, A., Angyan, J.G.: J. Chem. Phys. **132**, 244108(1–9) (2010)

Chapter 7
Orbital Energy

7.1 Koopmans Theorem

Orbital energies are the solutions of the Kohn–Sham and Hartree–Fock equations, which form the basis of electronic state theories. In the field of chemistry, orbitals have been employed as a significant tool for analysis. For example, frontier orbital theory (Fukui et al. 1952) uses the electronic distributions of the HOMO and LUMO to specify reaction sites and reactivities. Orbital energies were also used as important tools for analyzing reactions in early reaction theories, including the frontier orbital theory. Even now, orbital energies are often applied to the analyses of experimental results in considering reaction mechanisms. However, hypothetical orbital energies are employed instead of the calculated ones in such analyses. This is because *accurate orbital energies are not capable of being reproduced by the Kohn–Sham and Hartree–Fock equations*. Actually, *orbital energies have been considered skeptically, even for their physical meaning*. Nevertheless, orbitals have been physically established as the electronic motion states of systems. So, what do the corresponding orbital energies represent?

The physical meaning of orbital energies is clarified by *Koopmans theorem* (Koopmans 1934). For orbital ϕ_i, the orbital energy of the Hartree–Fock equation is represented by

$$\epsilon_i = \int d^3\mathbf{r}\phi_i^*(\mathbf{r})\hat{F}\phi_i(\mathbf{r}) = h_i + \sum_j^n \left(2J_{ij} - K_{ij}\right), \tag{7.1}$$

where h_i, J_{ij} and K_{ij} are the one-electron and two-electron integrals given in Eqs. (2.42) and (2.44), respectively. In this case, the total energy is provided as

$$E_0 = \sum_i^n h_i + \sum_{i<j}^n \left(2J_{ij} - K_{ij}\right) = \sum_i^n \epsilon_i - \sum_{i<j}^n \left(2J_{ij} - K_{ij}\right). \tag{7.2}$$

T. Tsuneda, *Density Functional Theory in Quantum Chemistry*,
DOI 10.1007/978-4-431-54825-6_7, © Springer Japan 2014

Moreover, the energy after removing one electron from orbital ϕ_i is derived as

$$E' = E_0 - h_i - \sum_j^n (2J_{ij} - K_{ij}) . \tag{7.3}$$

The ionization potential, which is the energy difference from E_0 to E', is, therefore, proven to be

$$\mathrm{IP} = E' - E_0 = -h_i - \sum_j^n (2J_{ij} - K_{ij}) = -\epsilon_i . \tag{7.4}$$

This indicates that the *occupied orbital energies are the corresponding negative of the ionization potentials*. This is the more narrowly defined version of Koopmans' theorem. Note that the proof of this theorem mentioned above neglects the SCF process, which is executed in standard calculations. In fact, *the Hartree–Fock SCF method violates the Koopmans theorem*. The calculated Hartree–Fock orbital energies resulting from SCF calculations are actually much more negative than the negative of the corresponding vertical ionization potentials, which are the ionization potentials for fixed structures of the systems.

It is easily proven that the Koopmans theorem is established for unoccupied orbitals (Szabo and Ostlund 1996). The energy after adding one electron to an unoccupied orbital ϕ_a is derived from Eq. (7.2) as

$$E'' = E_0 + h_a + \sum_j^n (2J_{aj} - K_{aj}) . \tag{7.5}$$

Therefore, the electron affinity, which is the energy difference from E'' to E_0, is proven to be

$$\mathrm{EA} = E_0 - E'' = -h_a - \sum_j^n (2J_{aj} - K_{aj}) = -\epsilon_a . \tag{7.6}$$

That is, the *unoccupied orbital energies are the corresponding negative of the electron affinities*. Differently from the occupied orbitals, the Hartree–Fock SCF method is found to give accurate LUMO energies, close to the negative of the corresponding vertical electron affinities, which again are the electron affinities for fixed structures. This may be because the effect of the SCF process, which is supposed to be an orbital relaxation effect, is small for unoccupied orbitals in Hartree–Fock calculations.

In connection with the Koopmans theorem, there is a method based on the *extended Koopmans theorem*, in which the ionization potentials of specific orbitals can be estimated without calculating the ionized electronic states, for which the electrons would have to be removed from the orbitals. The extended Koopmans theorem was independently proven by Day et al. (1974) and Morrell et al.

(1975), although Feynman (1954) also used a similar theorem in superfluid state calculations. In this method, vertical ionization potentials are calculated using the electron configuration energies after removing an electron from the occupied orbitals by solving

$$\mathbf{F}^{\mathrm{IP}} n = \boldsymbol{\Gamma} \mathbf{C} \iota, \tag{7.7}$$

where $\boldsymbol{\Gamma}$ is the metric matrix, in which the diagonal terms are the occupation numbers and the off-diagonal terms are zeros, and ι are the vertical ionization potentials. The transition matrix elements are given as

$$F_{ji}^{\mathrm{IP}} = \langle \Phi | \hat{a}_j^{\dagger} \left[\hat{H}, \hat{a}_i \right] | \Phi \rangle, \tag{7.8}$$

where $\hat{a}_i$ is the annihilation operator for an electron in the i-th orbital. Since this method gives the ionization potentials and electron affinities of specific orbitals (Piris et al. 2012) even for non-one-electron-SCF methods such as the MP2 method (Bozkaya 2013), the orbital energies of these orbital-free methods can be discussed virtually on the basis of the extended Koopmans theorem.

7.2 Janak's Theorem

Janak's theorem (Janak 1978) is also one of the most significant theorems having to do with orbital energy. It is noteworthy that *the Janak theorem is established not only for the Kohn–Sham and Hartree–Fock equations but also for overall one-electron SCF equations* in the form of Eq. (4.7). Let us consider the Kohn–Sham equation in Eqs. (4.6) and (4.10). Introducing the occupation numbers of orbitals, $\{n_i\}$, the Kohn–Sham equation is written as

$$\left(-\frac{1}{2}\nabla^2 + V_{\mathrm{ext}} + 2\sum_j^n \hat{J}_j + V_{\mathrm{xc}} \right) \phi_i = \epsilon_i \phi_i \tag{7.9}$$

$$E = T + E_{\mathrm{ext}}[\rho] + J[\rho] + E_{\mathrm{xc}}[\rho], \tag{7.10}$$

where T, E_{ext}, and J are the kinetic, external field, and Coulomb interaction energies, respectively, which are given by

$$T = \sum_i^n n_i \int d^3\mathbf{r} \phi_i^*(\mathbf{r}) \left(-\frac{1}{2}\nabla^2 \right) \phi_i(\mathbf{r}) = \sum_i n_i t_i, \tag{7.11}$$

$$E_{\mathrm{ext}}[\rho] = \int d^3\mathbf{r} \rho(\mathbf{r}) V_{\mathrm{ext}}, \tag{7.12}$$

and

$$J[\rho] = \frac{1}{2}\int d^3\mathbf{r}_1 d^3\mathbf{r}_2 \frac{\rho(\mathbf{r}_1)\rho(\mathbf{r}_2)}{r_{12}}. \tag{7.13}$$

In this case, the electron density, ρ, is represented by

$$\rho = \sum_i^n n_i |\phi_i|^2. \tag{7.14}$$

The variation of electronic energy E in terms of the occupation number is, therefore, given by

$$\frac{\partial E}{\partial n_i} = \frac{\partial T}{\partial n_i} + \frac{\partial (E_{\mathrm{ext}} + J + E_{\mathrm{xc}})}{\partial n_i} \cdot \frac{\partial \rho}{\partial n_i} \tag{7.15}$$

$$= t_i + 2\sum_j^n n_j \frac{\partial t_j}{\partial n_i}$$
$$+ \int d^3\mathbf{r} \left(2\sum_j^n \hat{J}_j + V_{\mathrm{xc}}\right) \cdot \left(|\phi_i|^2 + \sum_j^n n_j \frac{\partial |\phi_j|^2}{\partial n_i}\right). \tag{7.16}$$

Since using Eqs. (7.9) and (7.11),

$$t_i = \epsilon_i - \int d^3\mathbf{r} \left(2\sum_j^n \hat{J}_j + V_{\mathrm{xc}}\right) |\phi_i|^2. \tag{7.17}$$

is provided, substituting this into Eq. (7.16) leads to

$$\frac{\partial E}{\partial n_i} = \epsilon_i + \sum_j^n n_j \left[\frac{\partial t_j}{\partial n_i} + \int d^3\mathbf{r} \left(2\sum_j^n \hat{J}_j + V_{\mathrm{xc}}\right) \frac{\partial |\phi_j|^2}{\partial n_i}\right]. \tag{7.18}$$

By substituting

$$\frac{\partial t_j}{\partial n_i} = \int d^3\mathbf{r} \frac{\partial \phi_j^*}{\partial n_i} \left(-\frac{1}{2}\nabla^2\right) \phi_j + \int d^3\mathbf{r} \phi_j^* \left(-\frac{1}{2}\nabla^2\right) \frac{\partial \phi_j^*}{\partial n_i} \tag{7.19}$$

from Eq. (7.11) into Eq. (7.18), the energy derivative in terms of the occupation number is derived for an arbitrary orbital ϕ_i as

$$\frac{\partial E}{\partial n_i} = \epsilon_i + \sum_j^n n_j \left[\int d^3\mathbf{r} \frac{\partial \phi_j^*}{\partial n_i} \left(-\frac{1}{2}\nabla^2 + 2\sum_j^n \hat{J}_j + V_{\mathrm{xc}} \right) \phi_j \right.$$
$$\left. + \int d^3\mathbf{r} \phi_j^* \left(-\frac{1}{2}\nabla^2 + 2\sum_j^n \hat{J}_j + V_{\mathrm{xc}} \right) \frac{\partial \phi_j}{\partial n_i} \right] \quad (7.20)$$

$$= \epsilon_i + \sum_j^n n_j \epsilon_j \left[\int d^3\mathbf{r} \frac{\partial}{\partial n_i} |\phi_i|^2 \right] \quad (7.21)$$

$$= \epsilon_i + \sum_j^n n_j \epsilon_j \left[\frac{\partial}{\partial n_i} \int d^3\mathbf{r}\, |\phi_i|^2 \right] = \epsilon_i . \quad (7.22)$$

Therefore, the relationship between the total electronic energy and orbital energy is proven to be

$$\frac{\partial E}{\partial n_i} = \epsilon_i . \quad (7.23)$$

This equation indicates that *the derivative of the total electronic energy with respect to the occupation number of an orbital is identical to the orbital energy*. This is called Janak's theorem. Numerically, the Janak theorem is easily confirmed by calculating the total electronic energies of systems with *fractionally occupied* electronic states. That is, the gradient of the total electronic energy as a function of the occupation number is equal to the outermost orbital energy for each method. It is interesting to note that the total electronic energy can be calculated using the outermost orbital energies on the basis of Eq. (7.23) as follows:

$$E = \sum_{i=1}^n \int_0^1 \epsilon_i^{\mathrm{outermost}}(n_i) dn_i . \quad (7.24)$$

This equation, called the *Slater–Janak theorem* (Slater 1978; Janak 1978), produces very accurate total electronic energies (Elkind and Staroverov 2012).

In addition to the Janak theorem, there is an additional theorem deeply related to orbital energies. Perdew et al. (1982) proved that the total electronic energy varies linearly as a function of its fractional occupation number, i.e.,

$$E\left(n + \frac{p}{q}\right) = \frac{p}{q} E(n+1) + \frac{q-p}{q} E(n). \quad (7.25)$$

For this theorem, which is called the *energy linearity theorem for fractional occupations*, Yang et al. also proved the same equation based on the concepts of size consistency (see Sect. 3.3) and the translational invariance of energy

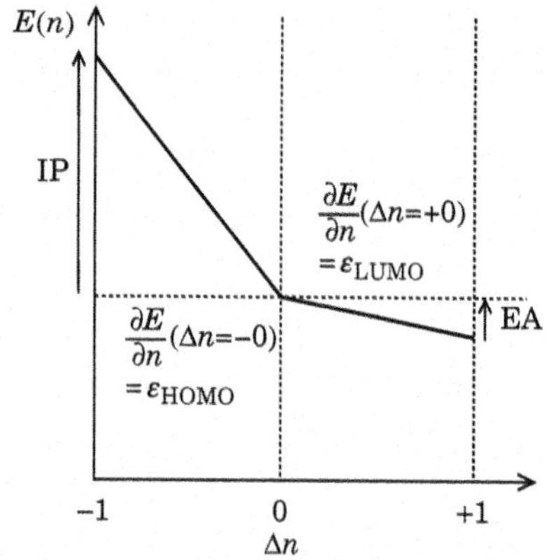

Fig. 7.1 Schematic diagram of total electronic energy as a function of fractional occupation number variation, Δn. Ionization potential (IP) and electron affinity (EA) are defined as $E(\Delta n = -1) - E(\Delta n = 0)$ and $E(\Delta n = 0) - E(\Delta n = +1)$, respectively. Based on the Janak theorem, the gradient of the total energy is HOMO energy for $\Delta n = -0$ and LUMO energy for $\Delta n = +0$. The linearly varied total energies also indicate that the outermost orbital energies are kept constant for the fractional occupation

(Yang et al. 2000). This theorem establishes the physical meaning of orbital energies by combining with the Janak theorem. In Fig. 7.1, a schematic diagram of the total electronic energy as a function of the fractional occupation number is illustrated. This figure clearly indicates that if the total electronic energy meets the energy linearity theorem, it is proven by the Janak theorem that the *HOMO and LUMO energies are identical to the corresponding negative of the ionization potential and the electron affinity, respectively* (Perdew et al. 1982). Since this is equivalent to the Koopmans theorem (see Sect. 7.1), this combined theorem is taken as *the Koopmans theorem for general one-electron SCF equations*. This energy linearity theorem has been used to test functionals for the reproducibilities of orbital energies, because it can be a requisite condition to reproduce correct orbital energies (see Sect. 7.9 for the concrete verification).

7.3 The Indispensability of Producing Accurate Orbital Energies

The conventional exchange-correlation functionals used in the Kohn–Sham equation contain electron correlations to some extent (see Sect. 4.5). Nevertheless, it has been recognized that the Kohn–Sham method is not able to reproduce accurate orbital energies. The cause for the inaccurate orbital energies has been investigated for many years in solid state physics from the viewpoint of the underestimation of band gaps. Let us consider the Kohn–Sham equation in Eq. (7.9). Perdew et al.

proved by the Janak and energy linearity theorems, as discussed in the previous section that *the underestimation of band gaps, which are assumed to be identical to the HOMO–LUMO gaps, can be attributed to the constant discontinuity of exchange-correlation potentials* (Perdew et al. 1982; Sham and Schlüter 1985),

$$V_{\mathrm{xc}}^{n+\Delta n} - V_{\mathrm{xc}}^{n-\Delta n} = \mathrm{const.} > 0, \tag{7.26}$$

where $V_{xc}^{n\pm\Delta n}$ is the exchange-correlation potential for occupation number $n \pm \Delta n$ ($\Delta n \to 0$). This discontinuity leads to the energy error (Perdew et al. 1982),

$$\Delta_{\mathrm{xc}} = \{\mathrm{IP} - \mathrm{EA}\} - \{\epsilon_{n+1}(n) - \epsilon_n(n)\}. \tag{7.27}$$

In this equation, IP and EA are the ionization potential and electron affinity, respectively, and $\epsilon_m(n)$ is the m-th orbital energy of the system containing n electrons. Sham and Schlüter (1985) established that *this energy error corresponds to the orbital energy variation for increasing numbers of electrons in the outermost orbitals*,

$$\Delta_{\mathrm{xc}} = \epsilon_{n+1}(n+1) - \epsilon_{n+1}(n). \tag{7.28}$$

This *outermost orbital energy invariance theorem* is often used as a criteria for evaluating the reproducibility of orbital energies (see Sect. 7.9). It is also proven that this energy error is related to the exchange-correlation potentials as

$$\Delta_{\mathrm{xc}} = \int d^3\mathbf{r} \left(V_{\mathrm{xc}}^{n+\Delta n} - V_{\mathrm{xc}}^{n-\Delta n}\right) \rho_{n+1}(\mathbf{r}). \tag{7.29}$$

Note that Eq. (7.28) is derived by assuming that the number of electrons n is sufficiently large, as seen in the electronic structures of solid state materials, i.e., $\epsilon_{n+1}(n+1) - \epsilon_{n+1}(n) = O(n^{-1}) \to 0$, in the original paper. However, this equation is easily proven by applying the Janak theorem to the energies of fractionally occupied systems.

Let us consider the exchange and correlation parts of the energy error separately. Perdew (1985) and Görling and Levy (1995) proved that *the energy error in the exchange part is attributable to the error only for the HOMO and LUMO*,

$$\Delta_{\mathrm{x}} = \int d^3\mathbf{r} d^3\mathbf{r}' \left[\phi_{n+1}^*(\mathbf{r})\phi_{n+1}(\mathbf{r}') - \phi_n^*(\mathbf{r})\phi_n(\mathbf{r}')\right] V_{\mathrm{x}}^{\mathrm{nl}}(\mathbf{r}, \mathbf{r}') - \int d^3\mathbf{r} \left[\rho_{n+1}(\mathbf{r}) - \rho_n(\mathbf{r})\right] V_{\mathrm{x}}(\mathbf{r}), \tag{7.30}$$

where $V_{\mathrm{x}}^{\mathrm{nl}}$ is the nonlocal exchange potential,

$$V_{\mathrm{x}}^{\mathrm{nl}}(\mathbf{r}) = \sum_i^n d^3\mathbf{r}' \phi_i^*(\mathbf{r}) \frac{1}{|\mathbf{r} - \mathbf{r}'|} \phi_i(\mathbf{r}'). \tag{7.31}$$

Görling and Levy (1995) also showed that *the energy error in the correlation part comes from the differences in the perturbations of n-electron systems with those of* $(n-1)$*- or* $(n+1)$*-electron systems and in correlation potential functionals at the* $\lambda \to \infty$ *limit of the uniform coordinate-scaling* (see Chap. 8) *for the HOMO and LUMO*, i.e.,

$$\Delta_c = \sum_I^{\text{SD exc.}} \frac{\left|\left\langle \Psi_{\text{KS}}(n+1) \left| \hat{V}_{\text{pert}} \right| \Psi_I(n+1) \right\rangle\right|^2}{E_{\text{KS}}(n+1) - E_I(n+1)}$$

$$+ \sum_I^{\text{SD exc.}} \frac{\left|\left\langle \Psi_{\text{KS}}(n-1) \left| \hat{V}_{\text{pert}} \right| \Psi_I(n-1) \right\rangle\right|^2}{E_{\text{KS}}(n-1) - E_I(n-1)}$$

$$-2 \sum_I^{\text{SD exc.}} \frac{\left|\left\langle \Psi_{\text{KS}}(n) \left| \hat{V}_{\text{pert}} \right| \Psi_I(n) \right\rangle\right|^2}{E_{\text{KS}}(n) - E_I(n)}$$

$$- \int d^3\mathbf{r} \left[\rho_{n+1}(\mathbf{r}) - \rho_n(\mathbf{r})\right] V_c^{\text{unif}}[\rho_n], \tag{7.32}$$

where $\Psi_{\text{KS}}(m)$ and $\Psi_I(m)$ indicate the ground electron configuration represented by the Slater determinant of the Kohn–Sham wavefunction of an m-electron system and its excited-state configuration, respectively, and $E_{\text{KS}}(m)$ and $E_I(m)$ are the corresponding total electronic energies. "SD exc." indicates the sum of singly-excited and doubly-excited configurations from the Kohn–Sham wavefunction. Moreover, the perturbation potential, $\hat{V}_{\text{pert}}$, and the correlation potential functional at the $\lambda \to \infty$ limit of the uniform coordinate scaling, V_c^{unif}, are given as

$$\hat{V}_{\text{pert}} = \hat{V}_{\text{ee}} - \left(2\sum_j^n \hat{J}_j + V_{\text{x}}\right), \tag{7.33}$$

and

$$V_c^{\text{unif}} = \lim_{\lambda \to \infty} V_c[\rho_\lambda], \tag{7.34}$$

respectively. What is significant is that *provided that a correlation functional satisfying the uniform coordinate scaling is used, the energy error coming from correlation functionals disappears* due to zero V_c^{unif} in Eq. (8.13). *The remaining perturbation terms are considered to come from the difference in the orbital relaxation effects due to the discrepancy in the number of electrons.* That is, these terms come from the disagreement of the configuration interactions with singly-excited electron configurations, because dynamical correlations, which account for the greater part of configuration interactions with doubly-excited configurations, must be included in the correlation functionals (see Sect. 4.5). Qian and Sahni

(2000) concluded that this energy error is attributable to the kinetic energy part of the correlation energies by considering the error for the Hartree–Fock approximation, in which no Δ_{x} is included. This supports the above explanation, because the kinetic energy part of the correlation energies is due mainly to orbital relaxation effects.

7.4 Electron Correlation Effects on Orbital Energies

Since the self-interaction error has been proven to cause the energy error in the exchange part, the poor quality of the orbital energies calculated by the Hartree–Fock method after the SCF process must be due to the insufficient electron correlations. Using Green function theory, Pickup and Goscinski determined *the effect of electron correlations on orbital energies* (Pickup and Goscinski 1973; Szabo and Ostlund 1996). According to this study, the difference of the orbital energy, ϕ_k, which is found as the pole of the Green function using the Fock operator, and the corresponding ionization potential for removing one electron from this orbital is given as

$$\begin{aligned}\Delta\epsilon_k &= \epsilon_k + \mathrm{IP} \\ &= \sum_{i\neq k}^{n_{\mathrm{occ}}}\sum_{a}^{n_{\mathrm{vir}}} \frac{|\langle ki|ka\rangle - \langle ki|ak\rangle|^2}{\epsilon_a - \epsilon_i} + \frac{1}{2}\sum_{i,j\neq k}^{n_{\mathrm{occ}}}\sum_{a}^{n_{\mathrm{vir}}} \frac{|\langle ij|ka\rangle - \langle ij|ak\rangle|^2}{\epsilon_a + \epsilon_k - \epsilon_i - \epsilon_j} \\ &\quad - \frac{1}{2}\sum_{i\neq k}^{n_{\mathrm{occ}}}\sum_{a,b}^{n_{\mathrm{vir}}} \frac{|\langle ab|ki\rangle - \langle ab|ik\rangle|^2}{\epsilon_a + \epsilon_b - \epsilon_i - \epsilon_k}. \end{aligned} \tag{7.35}$$

On the right-hand side of this equation, the first term sums the SCF-induced orbital relaxation effects of ionized molecules, the second term sums the electron-pair relaxation effects coming from the electron correlations between newly occupied molecular orbitals, $\{\phi_i\}$ and $\{\phi_j\}$, after the ionization and unoccupied orbitals, $\{\phi_k\}$, and the third term sums the electron-pair removal effects coming from the electron correlations between the electron pairs before the ionization and occupied orbitals, $\{\phi_k\}$ (Pickup and Goscinski 1973; Szabo and Ostlund 1996). That is, the discrepancy between the occupied Hartree–Fock orbital energies and the corresponding negative of the ionization potentials is concluded to result from the SCF-induced orbital relaxations and the differences of electron correlations (second-order perturbation effects) in the presence or absence of electrons, both of which are given as a result of the ionizations.

Pickup and Goscinski also proposed the cause for the poor quality of the unoccupied orbital energies of the Hartree–Fock method (Pickup and Goscinski 1973; Szabo and Ostlund 1996). For the case in which an electron occupies a virtual orbital, ϕ_c, the difference between its orbital energy, which is a pole of the Green function, and the corresponding negative of the electron affinity is given as

$$
\begin{aligned}
\Delta\epsilon_c &= \epsilon_c + \mathrm{EA} \\
&= \sum_{i}^{n_{\mathrm{occ}}} \sum_{a \neq c}^{n_{\mathrm{vir}}} \frac{|\langle ac|ci\rangle - \langle ac|ic\rangle|^2}{\epsilon_a - \epsilon_i} + \frac{1}{2} \sum_{i}^{n_{\mathrm{occ}}} \sum_{a,b \neq c}^{n_{\mathrm{vir}}} \frac{|\langle ab|ci\rangle - \langle ab|ic\rangle|^2}{\epsilon_a + \epsilon_b - \epsilon_i - \epsilon_c} \\
&\quad - \frac{1}{2} \sum_{i,j}^{n_{\mathrm{occ}}} \sum_{a \neq c}^{n_{\mathrm{vir}}} \frac{|\langle ij|ca\rangle - \langle ij|ac\rangle|^2}{\epsilon_a + \epsilon_c - \epsilon_i - \epsilon_j}.
\end{aligned} \tag{7.36}
$$

Similarly to the occupied orbitals, the first term of the right-hand side sums the orbital relaxation effects after the ionizations, the second term sums the electron-pair removal effects after the ionizations, and the third term sums the electron-pair relaxation effects before the ionizations (Pickup and Goscinski 1973; Szabo and Ostlund 1996). The discrepancy is, therefore, attributed to the SCF-induced orbital relaxations after the ionizations and the differences in the electron correlations (second-order perturbation effects) in the presence or absence of electrons.

Note, however, that by supplementing Eqs. (7.35) and (7.36), the orbital energies are improved to a lesser extent than expected in providing the corresponding negative of the ionization potentials and electron affinities. This indicates that the orbital energies cannot be accurately reproduced by adding the second-order perturbation-level electron correlations. That is, *highly sophisticated electron correlations are required to reproduce accurate orbital energies.*

7.5 Optimized Effective Potential Method

Thus far, various methods have been developed to produce correct orbital energies. The *optimized effective potential (OEP) method* is a representative example. In the OEP method, *orbital-dependent effective exchange-correlation potentials are obtained by solving an integral equation, which enables any potential to calculate the corresponding energy in a straightforward fashion.* As shown in Eq. (4.12), the expectation values of usual exchange-correlation potential functionals are not the corresponding exchange-correlation energies. Talman and Shadwick (1976) assumed that this inconsistency causes the poor quality of the orbital energies and developed the OEP method in order to obtain orbital-dependent effective exchange-correlation potentials. Similarly to the usual Kohn–Sham equation, the OEP equation is written as

$$
\left(-\frac{1}{2}\nabla^2 + V_{\mathrm{OEP}}\right)\phi_i = \epsilon_i\phi_i, \tag{7.37}
$$

$$
V_{\mathrm{OEP}} = V_{\mathrm{ext}} + 2\sum_{j}^{n} \hat{J}_j + V_{\mathrm{xc}}^{\mathrm{eff}}. \tag{7.38}
$$

In the OEP method, the effective exchange-correlation potential, $V_{\mathrm{xc}}^{\mathrm{eff}}$, is given by solving the integral equation,

$$t(\mathbf{r}) = \int d^3\mathbf{r}' \chi_{\mathrm{OEP}}(\mathbf{r}, \mathbf{r}') V_{\mathrm{xc}}^{\mathrm{eff}}(\mathbf{r}')$$

$$= \int d^3\mathbf{r}' \sum_i^{n_{\mathrm{occ}}} \frac{\delta E_{\mathrm{xc}}}{\delta \phi_i(\mathbf{r}')} \frac{\delta \phi_i(\mathbf{r}')}{\delta V_{\mathrm{OEP}}(\mathbf{r})}, \qquad (7.39)$$

$$\chi_{\mathrm{OEP}}(\mathbf{r}, \mathbf{r}') = \frac{\delta \rho(\mathbf{r})}{\delta V_{\mathrm{OEP}}(\mathbf{r}')} = 4 \sum_i^{n_{\mathrm{occ}}} \sum_a^{n_{\mathrm{vir}}} \frac{\phi_i(\mathbf{r})\phi_a(\mathbf{r})\phi_a(\mathbf{r}')\phi_i(\mathbf{r}')}{\epsilon_i - \epsilon_a}. \qquad (7.40)$$

This method is an extension of the Sharp–Hornton method for producing the local effective Hartree–Fock exchange potential (Sharp and Hornton 1953),

$$t(\mathbf{r}) = 4 \sum_i^{n_{\mathrm{occ}}} \sum_a^{n_{\mathrm{vir}}} \frac{\phi_i(\mathbf{r})\phi_a(\mathbf{r}) \int d^3\mathbf{r}' \phi_a(\mathbf{r}') V_{\mathrm{x}}^{\mathrm{HF}} \phi_i(\mathbf{r}')}{\epsilon_i - \epsilon_a}, \qquad (7.41)$$

$$V_{\mathrm{x}}^{\mathrm{HF}} = -\sum_j^{n_{\mathrm{occ}}} \frac{\phi_j(n)\phi_j(\mathbf{r}')}{|\mathbf{r}' - \mathbf{r}|}. \qquad (7.42)$$

For the OEP method, several practical problems have been reported. Two types of solutions are known to solve the OEP equation: the quadrature method using numerical grids and the analytical method using basis functions. The quadrature method is better suited for calculating spherical systems such as atoms. However, it is unsuited to calculate molecules and solids. On the other hand, although the analytical method is applicable to the calculations of molecules and solids, it usually cannot produce correct, convergent results. The *Krieger-Li-Iafrate (KLI) approximation* was suggested to solve these practical problems (Krieger et al. 1992). In the KLI approximation, exchange potentials are obtained by solving

$$V_{\mathrm{x}}^{\mathrm{KLI}} = V_{\mathrm{x}}^{\mathrm{Slater}} + 2 \sum_i^{n-1} \frac{\phi_i(\mathbf{r})\phi_i(\mathbf{r})}{\rho(\mathbf{r})} \int d^3\mathbf{r}' \phi_i(\mathbf{r}') \left(V_{\mathrm{x}}^{\mathrm{KLI}} - V_{\mathrm{x}}^{\mathrm{HF}}\right) \phi_i(\mathbf{r}'), \qquad (7.43)$$

$$V_{\mathrm{x}}^{\mathrm{Slater}} = 2 \sum_i^{n} \frac{\phi_i(\mathbf{r})\phi_i(\mathbf{r})}{\rho(\mathbf{r})} \int d^3\mathbf{r}' \frac{\phi_i(\mathbf{r}')\phi_i(\mathbf{r}')}{|\mathbf{r} - \mathbf{r}'|}, \qquad (7.44)$$

where the integral part on the right-hand side of Eq. (7.43) is calculated using

$$\sum_i^{n-1} \left(\delta_{ji} - M_{ji}\right) \int d^3\mathbf{r} \phi_i(n) \left(V_{\mathrm{x}}^{\mathrm{KLI}} - V_{\mathrm{x}}^{\mathrm{HF}}\right) \phi_i(\mathbf{r})$$

$$= \int d^3\mathbf{r} \phi_j(\mathbf{r}) \left(V_{\mathrm{x}}^{\mathrm{Slater}} - V_{\mathrm{x}}^{\mathrm{HF}}\right) \phi_j(\mathbf{r}) \ (j = 1, \cdots, n-1), \qquad (7.45)$$

$$M_{ij} = \int d^3\mathbf{r}\frac{\rho_j(\mathbf{r})\rho_i(\mathbf{r})}{\rho(\mathbf{r})}. \tag{7.46}$$

This integral equation simply produces the same effect as the localization of the nonlocal Hartree–Fock exchange potential.

Electron correlations are required to reproduce accurate orbital energies. *There are two approaches to incorporating electron correlations in the OEP–KLI method: constructing perturbation potentials and using correlation potential functionals* (Grabo and Gross 1996; Tong and Chu 1997). The details of these approaches have not been included in this book. In short, the difference between them is the correction for V_x^{HF} in Eq. (7.41) or Eq. (7.43), which is a perturbation potential in the former and a correlation potential functional in the latter. Moreover, the perturbation potential approach has two types, assuming the invariance of the perturbation for exchange-correlation potentials (Holleboom et al. 1988) and of the electron density corresponding to the exchange-correlation potentials (Sham and Schlüter 1983). There are only a few examples of orbital energy calculations using the OEP–KLI method. In particular, no orbital energy calculation has been carried out for the perturbation potential case. Table 7.1 summarizes some examples of calculated orbital energies compared to the corresponding negatives of the ionization potentials (Hamel et al. 2002; Kim et al. 1999). "$+\Delta$" indicates the addition of an energy shift to the potential. This table shows that *orbital energies are not quantitatively reproduced by LDA, even using an empirical energy shift*, and *the reproducibility worsens as the molecular size increases, even using the exact exchange potential* (see Sect. 7.6 for the details of this potential). Actually, there appears to be no example of the OEP method reproducing accurate orbital energies with no empirical correction. In addition, since the OEP method has the serious problems of poor SCF convergence and long computational time, the target systems of orbital energy calculations using this method have been restricted to atoms and small molecules.

7.6 Highly Correlated Correlation Potentials

As mentioned above, higher-order electron correlations are required in exchange-correlation potentials to reproduce correct orbital energies. *For incorporating higher-order electron correlations, one of the best strategies is to use the existing knowledge base on ab initio wavefunction theories.* A representative method following this strategy is the *ab initio DFT* (Bartlett et al. 2005a). Bartlett et al. developed this method setting five conditions: (1) all calculations are performed using the analytical integrations of basis functions as in ab initio methods, (2) a rigorous orbital-dependent exchange-correlation energy functional is taken from ab initio wave-function theories, (3) there is convergence to the right answer in the basis set and correlation limit, (4) the Slater determinant is consistent with the corresponding exchange-correlation potential functional, and (5) the exchange-correlation potentials are multiplicative but nonlocal.

Table 7.1 HOMO energies calculated by the OEP–KLI method compared to the corresponding negative ionization potentials in eV

Molecule	LDAXC	LDAXC +Δ	−IP	EXX LDAC	EXX PBEC	−IP Expt.
H_2	−11.2	−16.1	−16.2	−16.3	−15.8	−15.4
LiH	−2.3	−8.2	−7.7			−7.7
Li_2	−6.3	−5.0	−5.1			−5.0
Na_2	−0.3	−4.6				−4.9
K_2	−1.4	−3.6				−4.0
HF	−51.6	−17.7				−16.0
F_2	−10.1	−18.2	−17.4			−15.7
CO	−96.3	−15.1	−14.5	−16.6	−16.1	−14.0
N_2	−17.9	−17.2	−14.1	−18.8	−18.2	−15.6
P_2	−4.2	−10.1	−15.5			−10.5
H_2O	−23.8	−13.9		−15.2	−17.7	−12.6
CH_3	−9.2	−10.8	−13.7			
NH_3	−48.6	−13.2				−10.7
CH_4	−44.0	−14.8	−11.6	−16.3	−15.8	−14.3
CH_2O	16.1	−12.0	−14.1			−10.9
C_2H_4	7.8	−10.2				−10.7
C_6H_6	2.2	−11.3				
C_5H_5N	−0.5	−9.5				−9.6

The "+Δ" indicates the addition of an energy shift to the functional potential. Auxiliary basis functions for the density fitting technique (see Sect. 2.7) are used. Excerpt from Hamel et al. (2002) and Kim et al. (1999)

In the ab initio DFT, *the exact exchange (EXX) potential* and frequently its hybridization with the Hartree–Fock exchange potential is used as the exchange potential. The EXX potential that Görling and Levy developed is *a Kohn–Sham orbital-dependent potential* (Görling and Levy 1994; Ivanov et al. 1999),

$$V_{\rm x}(\mathbf{r}) = \frac{\delta E_{\rm x}[\rho(\mathbf{r})]}{\delta\rho(\mathbf{r})} = 4\sum_{i}^{n_{\rm occ}}\sum_{j}^{n_{\rm occ}}\sum_{a}^{n_{\rm vir}}\int d^3\mathbf{r}'\left[K_{ij}\frac{\phi_a(\mathbf{r}')\phi_i(\mathbf{r}')}{\epsilon_i-\epsilon_a}\right]\frac{\delta V_{\rm KS}(\mathbf{r}')}{\delta\rho(\mathbf{r})}, \quad (7.47)$$

where K_{ij} is the exchange energy in Eq. (7.31), in which the orbitals are the Kohn–Sham ones. In this equation, $\delta V_{\rm KS}/\delta\rho(\mathbf{r})$ is the inverse of the linear response function, $\chi_{\rm KS}$,

$$\chi_{\rm KS}(\mathbf{r},\mathbf{r}') = \frac{\delta\rho(\mathbf{r})}{\delta V_{\rm KS}(\mathbf{r}')} = 4\sum_{i}^{n_{\rm occ}}\sum_{a}^{n_{\rm vir}}\frac{\phi_i(\mathbf{r})\phi_a(\mathbf{r})\phi_a(\mathbf{r}')\phi_i(\mathbf{r}')}{\epsilon_i-\epsilon_a}. \quad (7.48)$$

This equation is based on the OEP method, as recognized from its similarity to Eq. (7.40). Even though the EXX potential produces valence orbital energies that are exactly similar to those of the Hartree–Fock exchange potential, it provides core orbital energies that are much different from the Hartree–Fock ones. Combining the EXX potential with a correlation potential mentioned later produces errors in the calculated core orbital energies that have opposite signs to the Hartree–Fock ones, crossing the exact values. Therefore, in ab initio DFT, the EXX potential is often mixed with the Hartree–Fock exchange potential, similarly to the hybrid functionals (see Sect. 5.5),

$$V_x = \kappa V_x^{\mathrm{HF}} + (1-\kappa) V_x^{\mathrm{EXX}}. \tag{7.49}$$

This hybridization drastically improves the results of core orbital energies.

For correlation potentials, the PT2H and PT2SC correlation potentials, which are *modified second-order perturbation potentials incorporating advanced electron correlations of the CCSDT method* (see Sect. 3.5), are often used (Schweigert and Bartlett 2008). The PT2H correlation potential is derived for the above-mentioned hybrid exchange potential from the following orbital-dependent energy expression (Schweigert et al. 2006),

$$E_c = \sum_i^{n_{\mathrm{occ}}} \sum_a^{n_{\mathrm{vir}}} t_i^a f_{ia} + \frac{1}{4} \sum_{i,j}^{n_{\mathrm{occ}}} \sum_{a,b}^{n_{\mathrm{vir}}} \left(t_{ij}^{ab} - t_i^a t_j^b - t_i^b t_j^a \right) \left(\langle ij|ab\rangle - \langle ij|ba\rangle \right), \tag{7.50}$$

$$t_i^a = \frac{K_{ia} - \int d^3\mathbf{r} \phi_i(\mathbf{r}) V_x^{\mathrm{EXX}} \phi_a(\mathbf{r})}{\epsilon_i - \epsilon_a} (1-\kappa), \tag{7.51}$$

$$t_{ij}^{ab} = \frac{\langle ij|ab\rangle - \langle ij|ba\rangle}{\epsilon_i + \epsilon_j - \epsilon_a - \epsilon_b}, \tag{7.52}$$

where $f_{ia} = \langle \Phi_{i\to a} | \hat{H} | \Phi_{\mathrm{KS}} \rangle$. Even though this f_{ia} is always zero in standard one-electron SCF methods containing the diagonalization of the Fock matrix according to the Brillouin theorem (see Sect. 3.4), it is not zero in the ab initio DFT due to the use of the Brueckner determinant (Brueckner 1954; Nesbet 1958), which has the maximum overlap with the Slater determinant for the ground state. Moreover, $\langle ij|ab\rangle$ is the integral in Eq. (4.42). The "SC" in the PT2SC correlation potential indicates "semicanonical." The PT2H potential uses the perturbation energy, including the first-order perturbation terms, which are given by the diagonalization of the Fock matrix, violating the Brillouin theorem. On the other hand, the PT2SC potential contains only the occupied–virtual orbital pairs in the perturbation terms by diagonalizing the Fock matrix separately for the occupied–occupied and virtual–virtual orbital pairs (Bartlett et al. 2005b). The difference between PT2H and PT2SC is confirmed to be negligible.

Using the exchange-correlation potential (HF + EXX + PT2SC) mentioned above, the valence occupied orbital energies are obtained with considerable accuracies for the first time. Table 7.2 displays the calculated results of orbital energies (Schweigert and Bartlett 2008). This method reproduces orbital energies with chemical accuracy: the mean absolute errors are less than 0.2 eV for valence

Table 7.2 Calculated occupied and virtual orbital energies of H_2 and CO molecules compared to the negative of the ionization potentials and electron affinities in eV

Orbital	HF+PT2H	EXX+PT2SC	HF+EXX+PT2SC	−IPExpt.
H_2O				
$1b_1$	−13.06 (−0.44)	−12.75 (−0.13)	−12.80 (−0.18)	−12.62
$1a_1$	−15.14 (−0.40)	−14.94 (−0.20)	−14.94 (−0.20)	−14.74
$1b_2$	−18.69 (−0.18)	−18.62 (−0.11)	−18.56 (−0.05)	−18.51
$2a_1$	−30.07 (2.54)	−34.54 (−1.93)	−32.21 (0.40)	−32.61
$1a_1$	−518.8 (20.9)	−562.3 (−22.6)	−540.6 (−0.9)	−539.7
CO				
5σ	−13.60 (0.41)	−13.63 (0.38)	−13.59 (0.42)	−14.01
1π	−17.48 (−0.57)	−16.75 (0.16)	−17.04 (−0.13)	−16.91
4σ	−18.83 (0.89)	−20.04 (−0.32)	−19.36 (0.36)	−19.72
3σ	−36.96 (1.34)	−41.98 (−3.68)	−39.37 (−1.07)	−38.30
2σ	−282.40 (13.80)	−313.10 (−16.90)	−297.80 (−1.60)	−296.20
1σ	−521.1 (21.5)	−565.0 (−22.4)	−543.1 (−0.5)	−542.6

The triple-zeta set of atomic natural orbitals are used as basis functions. Errors are shown in parentheses. Excerpt from Schweigert and Bartlett (2008)

occupied orbitals and less than 1 eV for core orbitals. Note, however, that no accurate core orbital energy is given using only the EXX potential without mixing with the Hartree–Fock exchange integral. This result clarifies the fact that *highly correlated potentials, at the level of coupled-cluster expansion theory, can give valence occupied orbital energies quantitatively, although even such highly correlated functionals cannot give core orbital energies accurately.*

7.7 Constrained Search for Exact Potentials

As another interesting method for obtaining highly correlated potentials, there is the constrained search method for determining potentials directly from the highly accurate electron densities of ab initio wavefunction theories, which is mentioned in Sect. 4.5. Wu and Yang developed *a method for producing kinetic-exchange-correlation potentials from accurate electron densities* by combining the OEP method for constructing local potentials from exchange-correlation energies (see Sect. 7.5) and the ZMP method for determining exchange-correlation potentials from highly accurate electron densities (see Sect. 4.5) (Wu and Yang 2003). In this method, the kinetic energy is obtained by maximizing

$$T_{\mathrm{WY}} = \sum_{i}^{n} \int d^3\mathbf{r}\phi_i(\mathbf{r}) \left(-\frac{1}{2}\nabla^2\right) \phi_a(\mathbf{r}) + \int d^3\mathbf{r} \left[\rho(\mathbf{r}) - \rho_{\mathrm{int}}(\mathbf{r})\right] V_{\mathrm{WY}}(\mathbf{r}), \tag{7.53}$$

where

$$V_{\mathrm{WY}}(\mathbf{r}) = V_{\mathrm{ext}}(\mathbf{r}) + \left(1 - \frac{1}{n}\right) \int d^3\mathbf{r}' \frac{\rho_0(\mathbf{r})}{|\mathbf{r} - \mathbf{r}'|} + \sum_p C_p \chi_p(\mathbf{r}), \quad (7.54)$$

and ρ_0 and ρ_{int} correspond to the calculated electron density of higher-level ab initio wavefunction theories and an initial electron density, respectively. The point is to represent the exchange-correlation potential as the linear combination of basis functions, χ_p, which are usually Gaussian-type basis functions (see Sect. 2.6). Consequently, the exchange-correlation potential is given as

$$V_{\mathrm{xc}}^{\mathrm{WY}}(\mathbf{r}) = \sum_p C_p \chi_p(\mathbf{r}) + \left(1 - \frac{1}{n}\right) \int d^3\mathbf{r}' \frac{\rho_0(\mathbf{r})}{|n - \mathbf{r}'|} - \int d^3\mathbf{r}' \frac{\rho(\mathbf{r})}{|\mathbf{r} - \mathbf{r}'|}. \quad (7.55)$$

Orbital energies can be calculated using the kinetic, exchange, and correlation potentials in Eqs. (7.53) and (7.55). Actually, HOMO energies are calculated to be close to the negative of the experimental ionization potentials for atoms and small molecules. Tozer et al. also calculated LUMO energies by combining this constrained search method with the condition on the energy error found in Eq. (7.27) (Teale et al. 2008). In Table 7.3, the HOMO and LUMO energies of small molecules, in which the "WY" results are calculated by the combined method, are excerpted. These calculations use the highly accurate electron densities determined by the CCSD(T) method (see Sect. 3.5). As shown in the table, accurate HOMO energies are obtained using the potentials determined directly from the highly accurate electron densities. This result supports the conclusion in the previous section that highly correlated potentials at the level of coupled-cluster theory can provide valence orbital energies quantitatively. However, these potentials give negative LUMO energies, which are clearly inconsistent with experimental results. The errors in the calculated LUMO energies are much larger even than those of typical pure functionals. Therefore, this indicates that *LUMO energies cannot be accurately reproduced, even using the potentials determined from highly correlated electron densities.*

7.8 Corrections for Orbital Energy Gaps in Solids

Thus far, in calculations of solid state systems, orbital energy calculations have been conventionally carried out as an approximation to the energy bands, which are then used to discuss photoexcitation. However, in this sense, the energy bands should be thought of as the excited states of solids, and therefore, the band gaps should be taken as the excitation energies of the solids and should be calculated with TDDFT. Nevertheless, the orbital energy gaps have been assumed to be identical to the excitation energies, without clear evidence. Based on this assumption, many

Table 7.3 Calculated HOMO and LUMO energies of small molecules compared to the corresponding negative of the ionization potentials and electron affinities in eV

Molecule	HOMO				LUMO			
	WY	PBEXC	B3LYP	−IP	WY	PBEXC	B3LYP	−EA
SO_2	−12.19	−8.00	−9.25	−12.49	−8.44	−4.41	−3.67	–
Cl_2	−11.48	−7.29	−8.38	−11.48	−8.25	−4.22	−3.32	–
F_2	−15.24	−9.44	−11.43	−15.70	−11.37	−5.80	−4.41	–
H_2CO	−10.86	−6.26	−7.56	−10.91	−6.94	−2.67	−1.66	1.50
C_2H_4	−10.67	−6.78	−7.56	−10.67	−4.87	−1.09	−0.24	1.80
CO	−13.96	−9.03	−10.42	−14.01	−6.72	−1.99	−1.03	1.80
PH_3	−10.23	−6.72	−7.59	−10.59	−3.65	−0.63	−0.38	1.90
H_2S	−10.42	−6.31	−7.24	−10.50	−4.52	−0.84	−0.52	2.10
HCN	−13.58	−9.03	−10.07	−13.61	−5.50	−1.09	−0.27	2.31
HCl	−12.74	−8.05	−9.14	−12.76	−5.28	−1.09	−0.68	3.29
CO_2	−13.63	−9.09	−10.37	−13.77	−4.90	−0.87	−0.52	3.81
NH_3	−10.83	−6.18	−7.40	−10.83	−4.52	−0.71	−0.46	5.61
HF	−15.95	−9.66	−11.46	−16.11	−5.80	−0.98	−0.65	5.99
H_2O	−12.60	−7.24	−8.73	−12.63	−5.20	−0.93	−0.63	6.39
CH_4	−14.29	−9.44	−10.69	−14.31	−4.30	−0.35	−0.19	7.81
MAE	0.11	4.79	3.54		8.87	4.79	4.29	

Auxiliary basis functions for expanding exchange-correlation potentials are used in the Wu-Yang (WY) constrained search method. Excerpt from Teale et al. (2008)

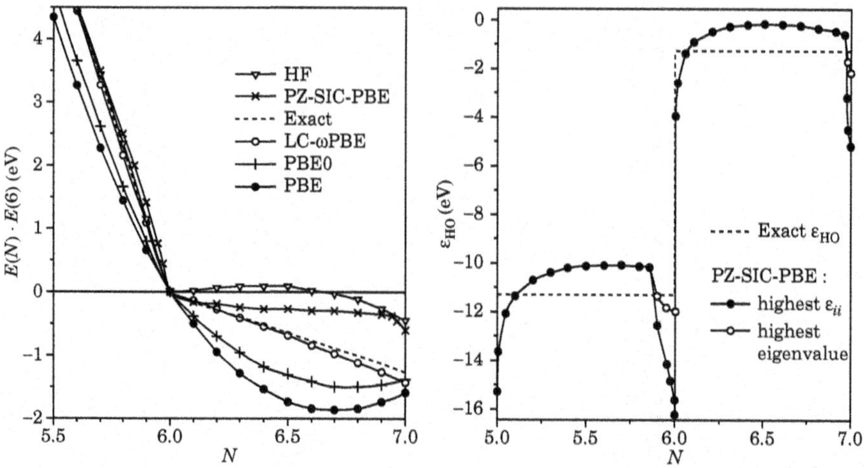

Fig. 7.2 Calculated total electronic energies (*left*) and orbital energies (*right*) of the carbon atom with respect to the fractional occupation number. The Perdew–Zunger (PZ) self-interaction corrected (SIC) PBE functional provides irregular behaviors for both the Kohn–Sham energies and orbital energies especially near integer occupation numbers. Reprinted from Vydrov et al. (2007)

studies have been carried out to attempt to make the orbital energy gaps between valence and conduction bands approach the experimental band gaps.

As mentioned in Sect. 7.3, the errors in exchange functionals have been accepted to cause the errors in orbital energies. Therefore, the self-interaction correction for exchange-correlation potential functionals has been examined with the aim of reproducing accurate orbital energies, as a natural result. However, conventional studies have not focused on the orbital energies of molecules for many years. Using the Perdew–Zunger self-interaction correction (see Sect. 6.2),

$$E = E_{\mathrm{KS}} - \sum_{i}^{n} (J_{ii} + E_{\mathrm{xc}}[\rho_i]) , \tag{7.56}$$

Vydrov et al. (2007) explored the reproducibility of Kohn–Sham orbital energies. Figure 7.2 shows the calculated total electronic energies and orbital energies of the self-interaction-corrected PBE functional with respect to the fractional occupation number of the outermost orbitals of the carbon atom to check whether this functional obeys the energy linearity theorem. In this figure, it is shown that the functional using this self-interaction correction meets neither the outermost orbital energy invariance theorem in Eq. (7.25) nor the energy linearity theorem in Eq. (7.28). Therefore, *the self-interaction corrected Kohn–Sham method does not give accurate orbital energies*. Note, however, that this method affects the GGA exchange-correlation functionals, which violate these theorems, in the direction of improvement, although these are overcorrected, so that there is no improvement in the behavior. This result clearly suggests that *the self-interaction error is not the main cause for the poor quality of the orbital energies, even restricted to the exchange functional part, although it may be partially responsible.*

Let us examine the self-energy correction based on the *GW approximation* as an approach in band calculations for reproducing accurate orbital energies. The self-energy, Σ, is defined as

$$\left(-\frac{1}{2}\nabla^2 + V_{\text{ext}} + 2\sum_j^n \hat{J}_j\right)\phi_i + \int d^3\mathbf{r}'\,\Sigma(\mathbf{r},\mathbf{r}';\epsilon_i)\phi_i(\mathbf{r}') = \epsilon_i\phi_i, \quad (7.57)$$

and it is described using the GW approximation (Hedin 1965; Szabo and Ostlund 1996) as

$$\Sigma(\mathbf{r},\mathbf{r}';\epsilon_i) = \frac{i}{2\pi}\int d\omega' G(\mathbf{r},\mathbf{r}';\omega-\omega')W(\mathbf{r},\mathbf{r}';\omega')\exp(-i\eta\omega'). \quad (7.58)$$

In Eq. (7.58), the *Green function*, G, is given as

$$G(\mathbf{r},\mathbf{r}';\omega) = \sum_k \frac{\phi_k(n)\phi_k^*(\mathbf{r}')}{\omega - \epsilon_k + i\eta}, \quad (7.59)$$

and W, the screened Coulomb interaction, is written as

$$W(\mathbf{r},\mathbf{r}';\omega) = \int d^3\mathbf{r}''\epsilon_{\text{de}}^{-1}(\mathbf{r},\mathbf{r}';\omega)v(\mathbf{r}''-\mathbf{r}'). \quad (7.60)$$

Moreover, ϵ_{de} is a frequency-dependent dielectric function and v, called the bond number, is a specific number of the calculated system. Although the details have not been included in this book, Johnson and Achcroft (1998) assumed that the energy error is attributable to the difference in the self-energy of exchange-correlation functionals from that of the Hartree–Fock exchange integral and suggested the form of the discontinuity, which is represented using the dielectric function for zero frequency, $\epsilon_{\text{de},0}$, and the electron density, for the LDA exchange-correlation potential,

$$\Delta V_{\text{xc}}^{\text{LDA}} = -\alpha\rho^{1/3}\left(\frac{1}{\epsilon_{\text{de},0}} - \frac{\beta\Delta\epsilon}{\epsilon_{\text{de},0}^2}\right), \quad (7.61)$$

where $\Delta\epsilon$ is the orbital energy gap between the valence and conduction bands and $\alpha = 1.14$ and $\beta = 6.12$ are semiempirical parameters. Using this correction, it has been reported that the orbital energy gaps of semiconductors approach the band gaps.

The *LDA+U method* (Liechtenstein et al. 1995) is one of the most frequently used corrections for orbital energy gaps in band calculations. In the LDA+U method, the difference of the Coulomb-exchange interactions from their averaged value is added to the total electronic energy of the LDA functional. Since this correction intends to incorporate spin-fluctuation effects, it is performed only for *d–d* and *f–f* orbital interactions. This correction has been reported to increase the orbital energy gaps of insulators by appropriate energies.

What should be noted is that, *since the corrections in the band calculations of solids are usually specialized to make the orbital energy gaps close to the band gaps, these are not available to improve the orbital energy calculations of molecules.*

7.9 Orbital Energy Reproduction by Long-Range Corrected DFT

Is it possible to develop a theory that accurately and comprehensively reproduces the orbital energies of molecules? Actually, there already exists a theory that quantitatively gives valence orbital energies. *The Kohn–Sham method using long-range corrected functionals* (see Sect. 6.1) *quantitatively reproduces valence orbital energies* (Tsuneda et al. 2010). Table 7.4 summarizes the calculated HOMO energies of atoms and small molecules compared to the corresponding negative ionization potentials. As clearly shown in the table, the long-range correction lowers the errors of the HOMO energies from the negative ionization potentials by an order of magnitude, to the level of quantitative values: the MAE values for LC-BOP are ca. 0.84 eV for typical molecules. Compared to the results of the pure (BOP) and hybrid (B3LYP) functionals, the outstanding accuracy of the LC functional becomes apparent. What is more important is the results of the LUMO energies. In Table 7.5, the calculated LUMO energies of atoms and small molecules are compared with the corresponding negatives of the electron affinities. This table indicates that the long-range correction drastically improves the LUMO energies, to an MAE of just 0.14 eV. Note that no additional semiempirical parameter is introduced in order to accurately reproduce the orbital energies. The original long-range corrected functionals contain only one parameter, μ, which is fixed at a value fitted for each functional that is corrected: e.g., for the B88 exchange functional, $\mu = 0.47$ is used for equilibrium electronic state calculations and $\mu = 0.33$ for response property calculations, including excited state calculations (see Sect. 6.1). In this chapter, it has thus far been described that no LUMO energy can be accurately reproduced, even using extremely highly correlated potential functionals. Furthermore, as far as the author knows, *even in advanced ab initio wavefunction methods incorporating higher-level electron correlations, there is no other theory that can quantitatively reproduce HOMO and LUMO energies simultaneously*. It is, therefore, surprising that highly accurate HOMO and LUMO energies are reproduced by the Kohn–Sham method only using long-range corrected functionals.

To confirm the reasonableness of the quantitative orbital energies of long-range corrected functionals, it is meaningful to examine the achievement of the energy linearity theorem, which is a necessary condition for providing correct

Table 7.4 Calculated HOMO energies of atoms and typical molecules and the errors from corresponding negative vertical ionization potentials in eV

	HF		BOP		B3LYP		LC-BOP		LC-PR-BOP	
System	ϵ_{HOMO}	Error	ϵ_{HOMO}	Error	ϵ_{HOMO}	Error	ϵ_{HOMO}	Error	ϵ_{HOMO}	Error
H	−13.60	0.00	−7.40	6.14	−8.68	4.89	−12.24	1.23	−13.60	0.00
He	−24.98	−1.52	−15.82	9.20	−17.90	6.95	−20.33	4.21	−25.81	−0.94
Ne	−23.15	−3.47	−13.24	8.38	−15.56	6.06	−18.02	3.85	−18.10	2.51
Ar	−16.08	−1.53	−10.08	5.50	−11.58	4.11	−14.33	1.46	−15.70	0.05
MAD		1.63		7.30		5.50		2.69		0.88
MA%D		8.65		39.04		29.57		4.65		4.48
C_2H_2	−11.23	−1.41	−6.97	4.28	−8.12	3.17	−10.87	0.56	−10.90	0.34
C_2H_4	−10.32	−1.37	−6.54	3.96	−7.59	2.92	−10.32	0.33	−10.36	0.00
CH_4	−14.81	−1.57	−9.30	4.54	−10.66	3.39	−13.36	0.81	−14.85	−0.47
Cl_2	−12.17	−1.00	−7.16	3.89	−8.45	2.95	−11.08	0.51	−12.14	−0.31
ClF	−13.48	−1.42	−7.70	4.59	−9.17	3.48	−11.85	0.88	−12.68	−0.04
CO	−15.11	−2.03	−8.92	4.93	−10.44	3.66	−13.22	1.00	−13.45	0.56
CO_2	−14.82	−2.42	−8.88	4.60	−10.37	3.35	−12.97	0.95	−12.98	0.45
CS	−12.57	−1.12	−7.24	3.97	−8.60	2.80	−11.28	0.24	−11.57	−0.39
F_2	−18.07	−2.02	−9.27	5.92	−11.35	4.41	−13.90	1.82	−13.97	1.26
H_2CO	−12.03	−2.56	−6.13	4.51	−7.57	3.22	−10.10	0.76	−10.45	0.28
H_2O	−13.88	−2.91	−7.09	5.60	−8.73	3.96	−11.46	1.44	−11.85	0.71
H_2S	−10.48	−1.28	−6.08	4.17	−7.22	3.11	−9.88	0.55	−10.59	−0.27
HCl	−12.97	−1.46	−7.82	4.76	−9.13	3.54	−11.85	0.91	−12.89	−0.31
HCOOH	−12.93	−2.92	−6.78	4.38	−8.30	3.07	−10.91	0.63	−11.05	0.24
HF	−17.69	−3.37	−9.51	6.73	−11.45	4.77	−14.09	2.37	−14.47	1.43
N_2	−16.79	−1.49	−10.15	5.16	−11.88	3.87	−14.64	1.26	−14.81	0.74
NH_3	−11.70	−2.30	−6.02	4.87	−7.42	3.49	−10.14	0.91	−10.47	0.43
P_2	−10.09	−0.91	−6.84	3.43	−7.75	2.59	−10.17	0.16	−10.44	0.09
PH_3	−10.57	−1.14	−6.53	3.86	−7.58	2.91	−10.12	0.39	−10.82	−0.40
SiH_3	−13.23	−1.28	−8.38	3.65	−9.57	2.79	−12.19	0.41	−13.54	−0.69
MAD		1.80		4.59		3.37		0.84		0.47
MA%D		15.98		37.52		27.18		6.38		3.57

Mean absolute deviations (MAD) and mean absolute percent deviations (MA%D) are also shown. The aug-cc-pVTZ basis functions are used. See Nakata and Tsuneda (2013)

Table 7.5 Calculated LUMO energies of atoms and typical molecules and the errors from corresponding negative vertical electron affinities in eV

	HF		BOP		B3LYP		LC-BOP		LC-PR-BOP	
System	ϵ_{LUMO}	Error	ϵ_{LUMO}	Error	ϵ_{LUMO}	Error	ϵ_{LUMO}	Error	ϵ_{LUMO}	Error
H	0.42	0.09	0.39	1.24	−0.03	0.81	0.37	1.07	0.42	0.96
He	2.70	0.00	1.36	−1.13	1.46	−0.98	2.64	0.00	2.66	0.00
Ne	5.47	0.04	2.54	−2.37	2.63	−2.14	4.88	−0.08	4.88	−0.08
Ar	2.76	0.03	0.70	−1.66	0.79	−1.45	2.47	0.01	2.61	−0.12
MAD		0.04		1.60		1.34		0.29		0.29
C_2H_2	0.80	0.00	0.14	−0.44	0.00	−0.60	0.77	0.01	0.78	0.01
C_2H_4	0.86	0.00	−0.18	−0.84	0.35	−0.42	0.80	0.00	0.82	0.00
CH_4	0.79	0.00	−0.29	−0.93	−0.19	−0.76	0.73	0.01	0.76	0.01
Cl_2	0.45	0.86	−4.29	−3.40	−3.62	−2.55	−1.25	−0.23	−1.09	−0.02
ClF	0.91	0.88	−4.36	−3.85	−3.56	−2.86	−1.10	−0.44	−0.93	−0.21
CO	1.80	0.01	−1.84	−3.01	−1.06	−2.22	1.23	−0.15	1.22	−0.43
CO_2	1.23	0.02	−0.68	−1.58	−0.51	−1.31	0.98	0.00	0.97	−0.01
CS	1.25	0.97	−3.21	−3.31	−2.56	−2.46	−0.21	−0.11	−0.22	−0.23
F_2	1.93	1.83	−5.35	−5.11	−4.16	−3.69	−1.71	−1.24	−1.71	−1.43
H_2CO	0.70	0.02	−0.35	−0.93	−0.24	−0.76	0.67	0.02	0.69	0.00
H_2O	0.79	0.01	−0.81	−1.40	−0.63	−1.14	0.63	0.00	0.69	−0.02
H_2S	0.76	0.02	−0.73	−1.26	−0.55	−1.01	0.63	0.02	0.68	−0.06
HCl	0.79	0.03	−1.01	−1.55	−0.70	−1.18	0.63	0.02	0.70	0.00
HCOOH	0.79	0.01	−0.51	−1.07	−0.35	−0.87	0.71	0.02	0.73	−0.01
HF	0.81	0.02	−0.86	−1.50	−0.66	−1.21	0.63	−0.01	0.70	−0.01
N_2	2.20	0.00	1.17	−0.52	1.26	−0.46	1.64	−0.43	1.61	−0.60
NH_3	0.80	0.01	−0.60	−1.20	−0.45	−0.99	0.67	0.01	0.73	0.02
P_2	0.41	0.64	−3.19	−2.85	−2.71	−2.17	−0.58	−0.01	−0.54	0.14
PH_3	0.71	0.01	−0.53	−1.05	−0.39	−0.85	0.61	0.01	0.66	−0.04
SiH_3	0.69	0.00	−0.37	−0.86	−0.24	−0.70	0.62	0.01	0.64	−0.05
MAD		0.27		1.83		1.41		0.14		0.17

Mean absolute deviations (MAD) are also shown. The aug-cc-pVTZ basis functions are used. See Nakata and Tsuneda (2013)

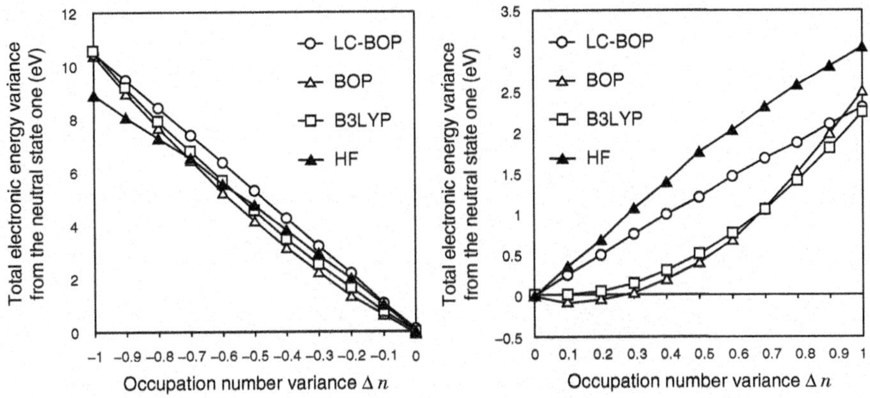

Fig. 7.3 Calculated total electronic energy of the ethylene molecule with respect to the variation of the occupation number of the outermost orbitals; the energy of the electroneutral state is set to be zero. The number of electrons decreases from the neutral state in the *left panel* and increases in the *right panel*. Calculated total electronic energies should vary linearly to give correct orbital energies. The aug-cc-pVQZ basis functions are used. See Tsuneda et al. (2010)

orbital energies (Yang et al. 2000). Figure 7.3 shows the total electronic energy of the ethylene molecule as a function of the degree of fractional occupation. In this figure, it is found that a linearly varying total electronic energy is produced, regardless of whether the number of electrons is increasing or decreasing, only for the use of a long-range corrected functional in the Kohn–Sham calculations. Meanwhile, pure and hybrid functionals give concave-shaped curves for the total electronic energies, independent of the increase or decrease of electrons. Although not mentioned in detail, it is numerically proven that the Janak theorem (see Sect. 7.2) is established for all one-electron SCF methods (Tsuneda et al. 2010). Since the gradients of the total electronic energies, which are discontinuous at the electroneutral state, are, therefore, equal to the HOMO and LUMO energies, *the concave-type variation of the total electronic energies for the fractional occupations indicates that there is an underestimation of the absolute values of the orbital energies*. In contrast, the absolute values of the orbital energies are overestimated by the Hartree–Fock method, which yields a convex-type variation of the total electronic energies. This discussion is consistent with the results of the calculated orbital energies in the Tables 7.4 and 7.5. Similar total electronic energy variances have been found for all of the systems thus far examined. That is, *only the long-range corrected functional meets the energy linearity theorem of total electronic energies for fractional occupations*, as expected.

The fractional occupation number dependence of the outermost orbital energies is also significant to test the validity of the orbital energies. As mentioned in Sect. 7.3, the outermost orbital energy invariance theorem should be satisfied in order to give correct orbital energies (Sham and Schlüter 1985). Figure 7.4 illustrates the fractional occupation number dependence of the outermost orbital energies

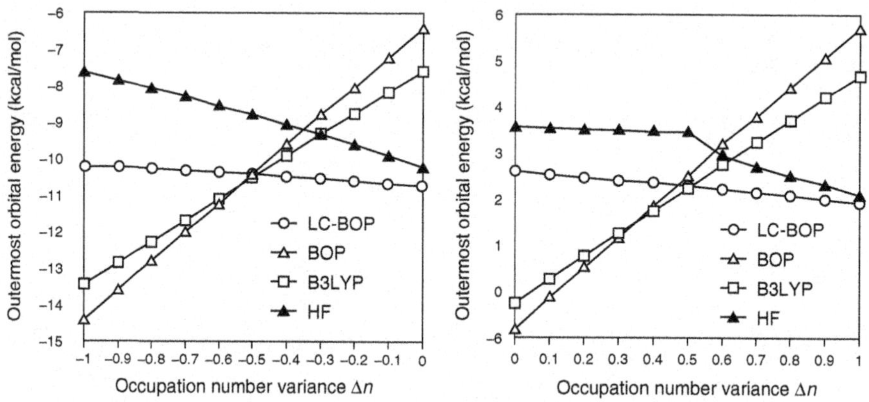

Fig. 7.4 Calculated outermost orbital energies of the ethylene molecule with respect to the fractional occupation number variance. The aug-cc-pVQZ basis functions are used. Calculated orbital energies should be kept constant to give correct orbital energies. See Tsuneda et al. (2010)

of ethylene. As shown in the figure, using a long-range corrected functional, the Kohn–Sham method gives a near-constant outermost orbital energy. That is, it nearly perfectly meets the outermost orbital energy invariance theorem. In contrast, the Kohn–Sham method using other functionals produces monotonically increasing orbital energies, and the Hartree–Fock method contrastingly produces monotonically decreasing orbital energies. Similar tendencies have also been found for other systems. This supports the idea that *accurate orbital energies of long-range corrected functionals are produced in a reasonable and versatile fashion.*

Why does the Kohn–Sham method accurately reproduce orbital energies only with the use of long-range corrected functionals? The cause is clarified by looking into the dependence of orbital energies on the occupation number. It is established that Kohn–Sham orbital energies have the dependence on the occupation number (Tsuneda et al. 2010),

$$\frac{\delta \epsilon_i}{\delta n_i} = \iint d^3\mathbf{r} d^3\mathbf{r}' \phi_i^*(\mathbf{r}) \phi_i^*(\mathbf{r}') \left[\frac{1}{|\mathbf{r} - \mathbf{r}'|} + \frac{\delta v_{\mathrm{xc}}}{\delta \rho} \right] \phi_i(\mathbf{r}) \phi_i(\mathbf{r}'). \tag{7.62}$$

That is, *the dependence of orbital energies on their occupation numbers is attributable to the self-interaction errors in the sum of the Coulomb self-interactions with the exchange-correlation self-interactions, through the integral kernels,* $f_{\mathrm{xc}} = \delta v_{\mathrm{xc}}/\delta \rho$. Therefore, *orbital energies cannot be correctly reproduced provided that self-interaction errors are included in the exchange integral kernel.* It may, however, be reasonable to consider that the inclusion of self-interaction errors in the exchange-correlation energies and potentials inevitably causes the errors in the exchange-correlation integral kernels and consequently makes it difficult to produce correct orbital energies. Figure 7.5 compares the energy distributions of the Coulomb self-interactions with those of the exchange-correlation

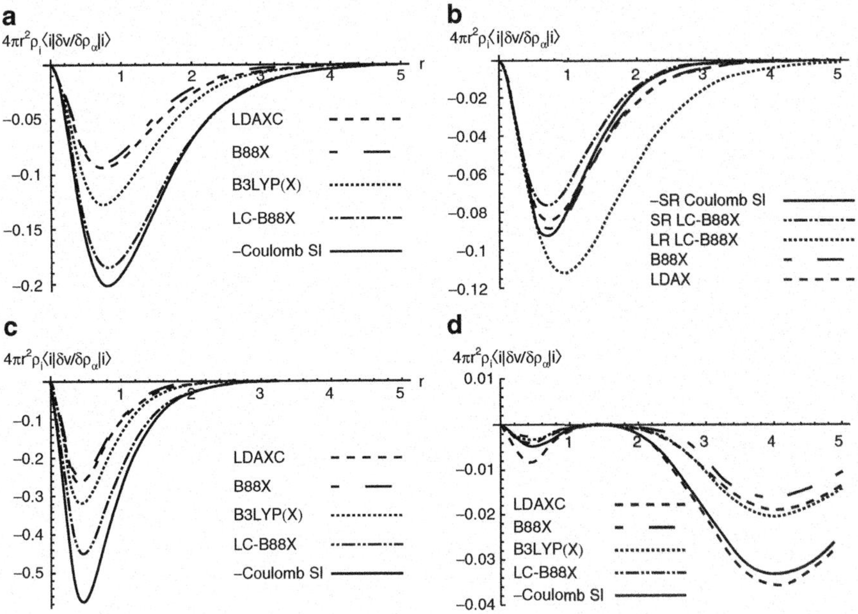

Fig. 7.5 Energy distributions of the Coulomb and exchange-correlation self-interactions through their integral kernels for (**a**) the HOMO of the hydrogen atom, (**b**) the components of (**a**), (**c**) the HOMO of the helium atom, and (**d**) the LUMO of the helium atom. See Tsuneda et al. (2010)

integral kernels through the integral kernels. For the HOMO of the hydrogen atom (Fig. 7.5a), the long-range correction clearly makes the self-interaction energy through the exchange-correlation integral kernel approach the Coulomb self-interaction energy, compared with the non-long-range corrected ones. Comparing the short-range and long-range component energies (Fig. 7.5b) shows that *the long-range component energy (LR) is much larger than the short-range one (SR) for the self-interaction through the exchange-correlation integral kernel.* Furthermore, by making a comparison between the results of the HOMO and LUMO of the helium atom (Fig. 7.5c and d), for which long-range corrected functionals give poor and accurate orbital energies, respectively, it is found that the self-interaction energy of the HOMO is insufficiently produced, even by a long-range corrected functional, while this functional produces a sufficiently large energy for that of the LUMO. The poor quality of the HOMO energy of the helium atom is, therefore, attributed to the considerable self-interaction error in the short-range part of the long-range corrected functional because of the large contribution of the short-range exchange part in this orbital.

What causes the poor-quality HOMO energies for hydrogen and the rare gases? As shown in Table 7.6, *similar problems are also found in the calculations of core orbital energies.* Note that even ab initio density functional theory using highly correlated correlation potentials also provides poor-quality core orbital energies

Table 7.6 Calculated core 1*s* energies of second- and third-row atoms in typical molecules and the errors from corresponding negative vertical ionization potentials in eV

	HF		BOP		B3LYP		LC-BOP		LC-PR-BOP	
System	ϵ_{1s}	Error	ϵ_{1s}	Error	ϵ_{1s}	Error	ϵ_{1s}	Error	ϵ_{1s}	Error
$\underline{C}H_4$	−304.98	−14.31	−268.97	21.60	−276.27	14.55	−273.37	16.32	−287.91	0.46
$Cl\underline{F}$	−717.19	−24.69	−662.16	31.12	−673.23	20.13	−666.76	25.87	−694.82	−3.89
$C\underline{O}$	−562.35	−20.98	−513.62	28.31	−523.48	18.61	−518.26	23.06	−542.85	−3.13
$\underline{C}O_2$	−311.86	−12.72	−274.31	22.65	−282.07	15.70	−279.27	17.52	−295.78	−0.07
$\underline{C}S$	−309.05	−14.76	−272.09	21.19	−279.64	14.04	−276.91	15.86	−295.52	−3.57
$H_2\underline{C}O$	−308.66	−14.69	−272.20	22.01	−279.65	14.92	−276.77	16.72	−292.82	−0.46
$H_2C\underline{O}$	−559.94	−21.84	−511.32	27.48	−521.13	17.77	−515.84	22.24	−539.98	−3.22
$H_2\underline{O}$	−559.62	−20.56	−510.91	28.37	−520.70	18.74	−515.40	23.15	−538.96	−1.80
$H\underline{C}OOH$	−310.11	−14.40	−273.23	21.93	−280.80	14.94	−277.97	16.77	−294.46	−0.73
$HC\underline{O}OH$	−561.27	−21.33	−512.50	27.50	−522.33	17.89	−517.03	22.32	−540.81	−2.81
$H\underline{F}$	−715.56	−22.58	−660.59	32.75	−671.62	21.86	−665.18	27.53	−692.75	−1.68
$\underline{N}H_3$	−422.90	−17.71	−380.47	24.74	−389.03	16.39	−384.89	19.51	−404.53	−1.42
MAD		18.38		25.80		17.13		20.57		1.94
MA%D		4.25		6.10		4.07		4.81		0.42
$H_2\underline{S}$	−2509.83	−31.09	−2403.73	71.12	−2424.65	50.83	−2408.22	65.91	−2479.68	−7.15
$H\underline{Cl}$	−2862.53	−32.13	−2749.14	76.94	−2771.49	55.29	−2753.67	71.75	−2834.42	−11.36
$\underline{P}H_3$	−2180.98	−29.67	−2082.17	65.66	−2101.68	46.74	−2086.70	60.41	−2157.19	−11.58
$\underline{Si}H_4$	−1875.63	−28.18	−1784.31	60.26	−1802.37	42.71	−1788.86	54.99	−1848.09	−5.26
$\underline{S}O_2$	−2515.61	−30.39	−2408.18	71.55	−2429.49	51.32	−2413.22	66.43	−2484.91	−6.60
MAD		30.29		69.11		49.38		63.90		8.39
MA%D		1.30		2.96		2.11		2.74		0.36

The 1*s* orbitals are positioned in underlined atoms. Mean absolute deviations (MAD) and mean absolute percent deviations (MA%D) are also shown. The aug-cc-pVTZ basis functions are used. See Nakata and Tsuneda (2013)

without mixing the Hartree–Fock exchange integral (see Sect. 7.6). Actually, it has been confirmed that *these poor-quality orbital energies, including the HOMO energies of hydrogen and the rare gases are drastically improved by a self-interaction-corrected functional.* This functional, called the LC-PR functional (Nakata and Tsuneda 2013; Nakata et al. 2010) (see Sect. 6.1), replaces the exchange energy in the self-interaction regions (see Sect. 6.2) with the exchange integral of the pseudospectral method only for the short-range part of the long-range corrected functionals. Tables 7.4 and 7.6 reveal that the core orbital energies and HOMO energies of hydrogen and the rare gases are quantitatively reproduced using this functional. What is important is that this functional maintains or even improves the accuracy of the molecular valence orbital energies of the long-range corrected functionals, as shown in Tables 7.4 and 7.5. It is, therefore, concluded that *the long-range correction is required to produce valence orbital energies quantitatively in the Kohn–Sham method, and an appropriate self-interaction correction is also required in order to reproduce core orbital energies and HOMO energies of rare gases accurately.*

References

Bartlett, R.J., Grabowski, I., Hirata, S., Ivanov, S.: J. Chem. Phys. **122**, 034104(1–12) (2005a)
Bartlett, R.J., Schweigert, I.V., Lotrich, V.F.: J. Chem. Phys. **123**, 062205(1–21) (2005b)
Bozkaya, U.: J. Chem. Phys. **139**, 154105(1–12) (2013)
Brueckner, K.A.: Phys. Rev. **96**, 508–516 (1954)
Day, O.W., Smith, D.W., Garrod, C.: Int. J. Quantum Chem. Symp. **8**, 501–509 (1974)
Elkind, P.D., Staroverov, V.N.: J. Chem. Phys. **136**, 124115(1–6) (2012)
Feynman, R.P.: Phys. Rev. **94**, 262–277 (1954)
Fukui, K., Yonezawa, T., Shingu, H.: J. Chem. Phys. **20**, 722–725 (1952)
Görling, A., Levy, M.: Phys. Rev. A **50**, 196–204 (1994)
Görling, A., Levy, M.: Phys. Rev. A **52**, 4493–4499 (1995)
Grabo, T., Gross, E.K.U.: Chem. Phys. Lett. **240**, 141–150 (1996)
Hamel, S., Casida, M.E., Salahub, D.R.: J. Chem. Phys. **116**, 8276–8291 (2002)
Hedin, L.: Phys. Rev. **139**, A796–A823 (1965)
Holleboom, L.J., Snijders, J.G., Baerends, E.J., Buijse, M.A.: J. Chem. Phys. **89**, 3638–3653 (1988)
Ivanov, S., Hirata, S., Bartlett, R.J.: Phys. Rev. Lett. **83**, 5455–5458 (1999)
Janak, J.F.: Phys. Rev. B **103**, 7165–7168 (1978)
Johnson, K.A., Ashcroft, N.W.: Phys. Rev. B **58**, 15548–15556 (1998)
Kim, Y.-H., Stadele, M., Martin, R.M.: Phys. Rev. A **60**, 3633–3640 (1999)
Koopmans, T.: Physica **1**, 104–113 (1934)
Krieger, J.B., Li, Y., Iafrate, G.J.: Phys. Rev. A **45**, 101–126 (1992)
Liechtenstein, A.I., Anisimov, V.I., Zaanen, J.: Phys. Rev. B **52**, 5467–5470 (1995)
Morrell, M.M., Parr, R.G., Levy, M.: J. Chem. Phys. **62**, 549–554 (1975)
Nakata, A., Tsuneda, T.: J. Chem. Phys. **139**, 064102(1–10) (2013)
Nakata, A., Tsuneda, T., Hirao, K.: J. Phys. Chem. A **114**, 8521–8528 (2010)
Nesbet, R.K.: Phys. Rev. **109**, 1632–1638 (1958)
Perdew, J.P.: In: Dreizler, R.M., da Providencia, J. (eds.) Density Functional Methods in Physics, Vol. 123 of NATO Advanced Science Institutes Series. Plenum, New York (1985)

Perdew, J.P., Parr, R.G., Levy, M., Balduz, J.L. Jr.: Phys. Rev. Lett. **49**, 1691–1694 (1982)
Pickup, B.T., Goscinski, O.: Mol. Phys. **26**, 1013–1035 (1973)
Piris, M., Matxain, J.M., Lopez, X., Ugalde, J.M.: J. Chem. Phys. **136**, 174116(1–6) (2012)
Qian, Z., Sahni, V.: Phys. Rev. B **62**, 16364–16369 (2000)
Schweigert, I.V., Bartlett, R.J.: J. Chem. Phys. **129**, 124109(1–8) (2008)
Schweigert, I.V., Lotrich, V.F., Bartlett, R.J.: J. Chem. Phys. **125**, 104108(1–14) (2006)
Sham, L.J., Schlüter, M.: Phys. Rev. Lett. **51**, 1888–1891 (1983)
Sham, L.J., Schlüter, M.: Phys. Rev. B **32**, 3883–3889 (1985)
Sharp, R.T., Hornton, G.K.: Phys. Rev. **90**, 317–317 (1953)
Slater, J.C.: The Self-Consistent Field for Molecules and Solids: Quantum Theory of Molecules and Solids. McGraw-Hill, New York (1978)
Szabo, A., Ostlund, N.S.: Modern Quantum Chemistry Introduction to Advanced Electronic Structure Theory. Dover, New York (1996)
Talman, J.D., Shadwick, W.F.: Phys. Rev. A **14**, 36–40 (1976)
Teale, A.M., De Proft, F., Tozer, D.J.: J. Chem. Phys. **129**, 044110(1–12) (2008)
Tong , X.-M., Chu, S.-I.: Phys. Rev. A **55**, 3406–3416 (1997)
Tsuneda, T., Song, J.-W., Suzuki, S., Hirao, K.: J. Chem. Phys. **133**, 174101(1–9) (2010)
Vydrov, O.A., Scuseria, G.E., Perdew, J.P.: J. Chem. Phys. **126**, 154109(1–9) (2007)
Wu, Q., Yang, W.: J. Chem. Phys. **118**, 2498–2509 (2003)
Yang, W., Zhang, Y., Ayers, P.W.:. Phys. Rev. Lett. **84**, 5172–5175 (2000)

Chapter 8
Appendix: Fundamental Conditions

Fundamental conditions indicate the conditions that should be met by the energy components such as the kinetic, exchange, and correlation energies. As described in Sect. 5.1, fundamental conditions have been used as significant criteria for assessing the physical validities of functionals developed. Actually, in solid state physics, in which extreme electronic states are often investigated, it appears to be conventional to use functionals that meet these conditions. In this chapter, let us briefly review major fundamental conditions.

1. Kinetic, exchange, and correlation energies have constant signs for nonzero electron density, $\rho \neq 0$, as follows:

$$T[\rho] > 0, \tag{8.1}$$

$$E_{\mathrm{x}}[\rho] < 0, \tag{8.2}$$

and

$$E_{\mathrm{c}}[\rho] \leq 0. \tag{8.3}$$

That is, for electrons, the *kinetic energies are positive definite* and the *exchange and correlation energies are negative definite*. Moreover, as far as an electron exists, the kinetic and exchange energies are nonzero due to the zero-point vibrational frequencies for kinetic energies and the self-interactions for exchange energies (see Sect. 6.2). In contrast, correlation energies are zero in one-electron systems. Note, however, that electron correlations are not necessarily zero for unoccupied orbitals, even in the hydrogen atom (Nakata and Tsuneda 2013).

2. In slowly varying electron density regions, where the density gradient $\nabla\rho$ is much smaller than the electron density ρ, the kinetic, exchange, and correlation energies are expanded using a dimensionless parameter $x_\sigma = |\nabla\rho_\sigma|/\rho_\sigma^{4/3}$ in Eq. (5.2) and $x = |\nabla\rho|/\rho^{4/3}$ as (von Weizsäcker 1935; Kleinman and Lee 1988)

T. Tsuneda, *Density Functional Theory in Quantum Chemistry*,
DOI 10.1007/978-4-431-54825-6_8,

$$\lim_{x_\sigma \to 0} T = \sum_\sigma \int d^3 n n \rho_\sigma^{5/3} \left[\frac{3}{5}(6\pi^2)^{2/3} + \frac{x_\sigma^2}{36} + O(x_\sigma^4) \right], \tag{8.4}$$

$$\lim_{x_\sigma \to 0} E_{\mathrm{x}} = -3\left(\frac{3}{4\pi}\right)^{1/3} \sum_\sigma \int d^3 \mathbf{r} \rho_\sigma^{4/3} \left[1 + \frac{5x_\sigma^2}{81(6\pi^2)^{2/3}} + O(x_\sigma^4) \right], \tag{8.5}$$

and

$$\lim_{x \to 0} E_{\mathrm{c}} = \int d^3\mathbf{r} \left\{ c_1[\rho] + c_2[\rho]x^2 + O(x^4) \right\}. \tag{8.6}$$

These equations are called the *generalized-gradient-approximation (GGA) limit conditions*, and, in particular, the $x_\sigma = 0$ values of these energies are called the *local density approximation (LDA) limit conditions*. Here, it should be noted that the coefficient of the x_σ^2 term in the exchange energy expansion in Eq. (8.5) is twice the conventional value (Kleinman and Lee 1988). The reason for this difference is mentioned later.

3. For rapidly varying (*high-density-gradient-low-density*) electron density regions, where the density gradient is much larger than the electron density, the kinetic and correlation energies behave as (Ma and Brueckner 1968; Dreizler and Gross 1990)

$$\lim_{x_\sigma \to \infty} T = \frac{1}{4} \sum_\sigma \int d^3\mathbf{r} \rho_\sigma^{5/3} x_\sigma^2, \tag{8.7}$$

and

$$\lim_{x \to \infty} \rho^{-1} \bar{E}_{\mathrm{c}} = 0, \tag{8.8}$$

where $\bar{E}_{\mathrm{c}}$ is the integral kernel of the correlation energy. It is interesting to note that the right-hand side of Eq. (8.7) is equivalent to the von Weizsäcker kinetic energy in Eq. (4.4). This indicates that the kinetic energy at the low-density-high-gradient limit is the von Weizsäcker kinetic energy. Moreover, the lack of a condition for the exchange energy at this limit has led to the development of various GGA exchange functionals, which are much different for large x_σ (see Sect. 5.2). This limit appears prominently in small electron density regions. In such regions, dispersion forces usually determine the character of the interatomic bonds. Actually, van der Waals bonds cannot be accurately reproduced using correlation functionals that violate Eq. (8.8). For example, the LYP correlation functional, which violates this condition, overestimates the correlation energies in van der Waals calculations of rare gas dimers (see Sect. 6.3) (Kamiya et al. 2002).

4. *Coordinate-scaling conditions* are also often used in fundamental conditions. In these conditions, the order of each energy component is regulated for the scaling of the electron density, which corresponds to the scaling of coordinates (see references in Tsuneda et al. 2001). There is the uniform coordinate-scaling for cubic coordinates (x, y, z), the second-order nonuniform coordinate scaling for planar coordinates (x, y), and the first-order nonuniform coordinate scaling for linear coordinates x. The following are established for uniform coordinate-scaling ($\rho(x, y, z) \rightarrow \rho_\lambda = \lambda^3 \rho(\lambda x, \lambda y, \lambda z)$):

$$T[\rho_\lambda] = \lambda^2 T[\rho], \tag{8.9}$$

$$E_{\mathrm{x}}[\rho_\lambda] = \lambda E_{\mathrm{x}}[\rho], \tag{8.10}$$

$$E_{\mathrm{c}}[\rho_\lambda] < \lambda E_{\mathrm{c}}[\rho] \ \ (\lambda < 1), \tag{8.11}$$

$$E_{\mathrm{c}}[\rho_\lambda] > \lambda E_{\mathrm{c}}[\rho] \ \ (\lambda > 1), \tag{8.12}$$

$$\lim_{\lambda \to \infty} E_{\mathrm{c}}[\rho_\lambda] = \text{const.} \neq 0, \tag{8.13}$$

and

$$\lim_{\lambda \to 0} \frac{1}{\lambda} E_{\mathrm{c}}[\rho_\lambda] = \text{const.} \neq 0. \tag{8.14}$$

For the nonuniform coordinate-scaling, the following conditions are also established: in the second-order scaling ($\rho(x, y, z) \rightarrow \rho_{\lambda\lambda}^{xy} = \lambda^2 \rho(\lambda x, \lambda y, z)$):

$$\lim_{\lambda \to \infty} \frac{1}{\lambda} E_{\mathrm{x}}[\rho_{\lambda\lambda}^{xy}] = \text{const.} \neq 0, \tag{8.15}$$

$$\lim_{\lambda \to 0} \frac{1}{\lambda} E_{\mathrm{x}}[\rho_{\lambda\lambda}^{xy}] = \text{const.} \neq 0, \tag{8.16}$$

$$\lim_{\lambda \to \infty} E_{\mathrm{c}}[\rho_{\lambda\lambda}^{xy}] = 0, \tag{8.17}$$

and

$$\lim_{\lambda \to 0} \frac{1}{\lambda^2} E_{\mathrm{c}}[\rho_{\lambda\lambda}^{xy}] = \text{const.} \neq 0, \tag{8.18}$$

and in the first-order scaling ($\rho(x, y, z) \rightarrow \rho_\lambda^x = \lambda \rho(\lambda x, y, z)$):

$$\lim_{\lambda \to \infty} E_{\mathrm{x}}[\rho_\lambda^x] = \text{const.} \neq 0, \tag{8.19}$$

$$\lim_{\lambda \to 0} E_{\mathrm{x}}[\rho_\lambda^x] = \text{const.} \neq 0, \tag{8.20}$$

$$\lim_{\lambda \to \infty} \lambda E_{\mathrm{c}}[\rho_\lambda^x] = \text{const.} \neq 0, \tag{8.21}$$

and

$$\lim_{\lambda \to 0} \frac{1}{\lambda} E_c[\rho_\lambda^x] = 0. \tag{8.22}$$

There is no nonuniform coordinate scaling condition for kinetic energy. It is interesting to evaluate the density dependence of the exchange and correlation energies at the scaling limit. For the exchange energy, the density dependences are given for three forms of density as

- $O(\rho)$ for linearly scaled density,
- $O(\rho^{3/2})$ for planarly scaled density, and
- $O(\rho^{4/3})$ for spherically scaled density.

A hypothesis is proposed by comparing these density dependences with the electron number dependences of the Hartree–Fock exchange integral given in conventional linear-scaling calculations. By lengthening a linear alkane, the computational time of the exchange integral calculations linearly increase with the number of electrons (Lambrecht and Ochsenfeld 2005). On the other hand, the electron number dependence of the computational time deviates significantly from linearity for the planar extension of a graphene sheet (Schwegler and Challacombe 1999). Interestingly, extending the size of a water cluster spherically leads to computational times that are slightly closer to linearity than that of graphene (Schwegler and Challacombe 1999). Comparing these with the density dependences of the exchange energy mentioned above, it is found that *the computational time of the exchange integral correlates with the coordinate scalings of the exchange energy*. Similarly, the density dependences of the correlation energy are provided as

- $O(1)$ for linearly extended density,
  $O(\rho^n)$ $(n > 2)$ for linearly contracted density,
- $O(\rho^m)$ $(m < 1)$ for planarly extended density,
  $O(\rho^2)$ for planarly contracted density,
- $O(\rho)$ for spherically extended density, and
  $O(\rho^{4/3})$ for spherically contracted density.

By comparing these with the density dependences of the exchange energy, the ratio of the correlation energy to the exchange energy can be evaluated for expanding systems. Lengthening calculated systems linearly leads to a ratio approaching $O(\rho^{-1/3})$, while extending systems planarly or spherically leads to a ratio approaching $O(\rho^{-1})$. This suggests that the correlation energy becomes less significant as the electron density increases in large systems, but it has a different significance for linear molecules and others.

5. As another significant fundamental condition, there is a condition on the self-interaction error. The self-interaction error is the Coulomb self-interaction, which should inherently cancel with the exchange self-interaction but remains due to the use of the exchange functional (see Sect. 6.2). Since one-electron systems

contain only the exchange self-interaction in two-electron interactions, *the self-interaction-free conditions for one-electron systems* have been suggested for determining whether or not the kinetic, exchange, and correlation energies are self-interaction-free (Zhang and Yang 1998):

$$T[q\rho_1] = O(q), \tag{8.23}$$

$$E_x[q\rho_1] = q^2 E_x[\rho_1], \tag{8.24}$$

and

$$E_c^{\alpha\beta}[q\rho_1] = 0, \tag{8.25}$$

where ρ_1 is the electron density of one-electron systems. Interestingly, Eq. (8.25) is derived from the density matrix of self-interacting electrons, similarly to *the far-from-nucleus (long-range) asymptotic behavior condition* for exchange energy (Levy et al. 1984),

$$\lim_{r\to\infty} \rho^{-1}\bar{E}_x = -\frac{1}{2r} \tag{8.26}$$

(see Sect. 6.2). For the same reason, the kinetic energy becomes the Weizsäcker one in Eq. (8.7) for one-electron systems. Therefore, the self-interaction error leads to the fact that most GGA exchange functionals violate the far-from-nucleus asymptotic behavior condition. There is also *the self-interaction-free condition for N electrons* (Mori-Sanchez et al. 2006). In this condition, the energy linearity theorem for fractional occupations (see Sect. 7.2) is used as the criterion. Note that long-range-corrected functionals meet this condition. Instead, there is thus far no example that this condition is met without the long-range correction (see Sect. 7.9). This suggests that *the long-range correction is required to remove the self-interaction errors from multiple electron systems*.

6. In addition, the *Lieb–Oxford bound condition* (Lieb and Oxford 1981),

$$E_x[\rho] \geq -1.679 \int d^3\mathbf{r}\rho^{4/3}. \tag{8.27}$$

has been often used to evaluate the validity of exchange functionals. Although this condition sets the upper limit of exchange energy, it is applicable only to GGA exchange functionals. Actually, the Hartree–Fock exchange integral violates this condition. Using the density functional prefactor, $\beta[\rho]$, a GGA correlation functional is also established to be the difference of the LDA exchange functional, $E_x^{\rm LDA}$, and the exact exchange energy, $E_x^{\rm exact}$, on the basis of this condition as follows (Odashima and Capelle 2009):

$$E_c[\rho] = \beta[\rho]\left(\lambda E_x^{\rm LDA} - E_x^{\rm exact}\right), \tag{8.28}$$

Table 8.1 Fundamental conditions for exchange energy and the comparison of GGA exchange functionals for their validities

Condition	LDA	PW91	PBE	B88	PF_{TFW}
Negative exchange energy	Yes	Yes	Yes	Yes	Yes
LDA limit	Yes	Yes	Yes	Yes	Yes
Slowly varying density limit	–	Yes?	Yes?	No	Yes?
Uniform coordinate scaling	Yes	Yes	Yes	Yes	Yes
Nonuniform coordinate scaling	No	No	No	No	No
Lieb–Oxford bound	Yes	Yes	Yes	No	Yes
Self-interaction-free for 1 electron	No	No	No	No	No
Far-from-nucleus asymptotic behavior	No	No	No	Y/N	No

Table 8.2 Fundamental conditions for the correlation energy and validity comparison of GGA correlation functionals

Condition	LDA	PW91	PBE	LYP	OP_{B88}
Negative correlation energy	Yes	No	Yes	No	Yes
LDA limit	No	No	No	No	Yes
Slowly varying density limit	–	Yes	Yes	No	Yes
Rapidly varying density limit	No	Yes	Yes	No	Yes
Uniform coordinate scaling	No	No	Yes	No	Yes
Nonuniform coordinate scaling	No	Yes	No	No	No
Self-interaction-free for 1 electron	No	No	No	Yes	Yes

where $\lambda = 1.679/\left((3/4)(3/\pi)^{1/3}\right) = 2.273$. Since this equation is trivial for an arbitrary $\beta[\rho]$, it is given using the LDA and GGA limit as

$$E_c[\rho] = \frac{E_c^{GGA}[\rho]}{\lambda E_x^{LDA}[\rho] - E_x^{GGA}[\rho]} \left(\lambda E_x^{LDA}[\rho] - E_x^{exact}\right), \tag{8.29}$$

where $E_x^{GGA}[\rho]$ is a GGA exchange functional satisfying Eq. (8.5) and $E_c^{GGA}[\rho]$ is a GGA correlation functional satisfying Eq. (8.6). Note that conventional correlation functionals violate this condition (Haunschild et al. 2012). However, this is still controversial, because the exact exchange energy, which violates Eq. (8.27), is applied to the condition for correlation functionals.

Next, let us examine to what extent conventional GGA functionals meet the fundamental conditions mentioned above. Tables 8.1 and 8.2 summarize the fundamental conditions of exchange and correlation energies, respectively, to show which conditions are obeyed by various GGA exchange and correlation functionals.

As shown in Table 8.1, GGA exchange functionals meet the fundamental conditions equivalently. PF_{TFW} indicates the parameter-free (PF) exchange functional in Eq. (5.10), in which the Thomas–Fermi–Weizsäcker (TFW) GGA kinetic energy functional is applied to the kinetic energy density part. This table shows that all of the GGA exchange functionals examined meet both the negative exchange energy and LDA limit conditions, while only the B88 exchange functional violates the slowly varying density limit and the Lieb–Oxford bound conditions. For the coordinate-scaling conditions, all of these functionals meet the uniform conditions but violate the nonuniform ones. The far-from-nucleus asymptotic behavior condition is met only for the Slater-type wavefunctions by the B88 functional. What should be noticed is the question mark for the slowly varying density limit condition. In the conventional slowly varying limit condition, the coefficient of x_σ^2 is $5/[162(6\pi^2)^{2/3}]$ (Kleinman and Lee 1988), which is just half of the coefficient in Eq. (8.5). This is because the coefficient of the PF_{TFW} is proven to be just twice the original one (Tsuneda and Hirao 2000). Since the original coefficient is actually known to be too small to use in practical calculations, conventional GGA exchange functionals have been modified to use approximately doubled coefficients: e.g., 0.003612, which is 1.78 times the original one, in the PBE exchange functional. Therefore, the correct coefficient must be $5/[81(6\pi^2)^{2/3}]$. However, the question mark is appended to the "Yes," because it has not yet been proven definitely.

Finally, let us consider the fundamental conditions for the correlation energy. OP_{B88} is the OP correlation functional in Eq. (5.30) using the B88 exchange functional in the exchange functional part in Eq. (5.31). For the correlation energy, the PW91 and LYP functionals violate even the negative correlation energy condition. Surprisingly, even major the LDA correlation functionals, the VWN LDA and the PW LDA functionals, violate the LDA limit condition. The PW91 and PBE functionals containing the PW LDA functional as the LDA limit, therefore, also disobey this condition. In contrast, the OP functional, containing no LDA functional, meets these conditions. Only the LYP functional violates the slowly varying and rapidly varying density limit conditions in GGA functionals. For coordinate-scaling conditions, only the PBE and OP functionals meet the uniform condition, and only the PW91 functional obeys the nonuniform ones. The self-interaction-free condition is met only by the LYP and OP functionals, which are both Colle–Salvetti-type functionals. What should be emphasized is the high physical validity of the OP functional, which meets all conditions except for the nonuniform coordinate-scaling conditions. Considering that the OP functional was developed without taking any fundamental conditions into account, its physical validity must have important implications.

References

Dreizler, R.M., Gross, E.K.U.: Density-Functional Theory An Approach to the Quantum Many-Body Problem. Springer, Berlin (1990)

Haunschild, R., Odashima, M.M., Scuseria, G.E., Perdew, J.P., Capelle, K.: J. Chem. Phys. **136**, 184102(1–7) (2012)

Kamiya, M., Tsuneda, T., Hirao, K.: J. Chem. Phys. **117**, 6010–6015 (2002)

Kleinman, L., Lee, S.: Phys. Rev. B **37**, 4634–4636 (1988)

Lambrecht, D.S., Ochsenfeld, C.: J. Chem. Phys. **123**, 184101(1–14) (2005)

Levy, M., Perdew, J.P., Sahni, V.: Phys. Rev. A **30**, 2745–2748 (1984)

Lieb, E.H., Oxford, S.: Int. J. Quantum Chem. **19**, 427–439 (1981)

Ma, S.K., Brueckner, K.A.: Phys. Rev. **165**, 18–31 (1968)
Mori-Sanchez, P., Cohen, A., Yang, W.: J. Chem. Phys. **125**, 201102(1–4) (2006)
Nakata, A., Tsuneda, T.: J. Chem. Phys. **139**, 064102(1–10) (2013)
Odashima, M.M., Capelle, K.: Phys. Rev. A **79**, 062515(1–6) (2009)
Schwegler, E., Challacombe, M.: J. Chem. Phys. **106**, 6223–6229 (1999)
Tsuneda, T., Hirao, K.: Phys. Rev. B **62**, 15527–15531 (2000)
Tsuneda, T., Kamiya, M., Morinaga, N., Hirao, K.: J. Chem. Phys. **114**, 6505–6513 (2001)
von Weizsäcker, C.F.: Z. Phys. **96**, 431–458 (1935)
Zhang, Y., Yang, W.: Phys. Rev. Lett. **80**, 890–890 (1998)

Index

T. Tsuneda, *Density Functional Theory in Quantum Chemistry*,
DOI 10.1007/978-4-431-54825-6, © Springer Japan 2014

H

I

J

K

L

M

N

O

P

R

S

T

U

V

W

X

Z

The manufacturer's authorised representative in the EU is Springer Nature Customer Service Centre GmbH, Europaplatz 3, 69115 Heidelberg, Germany. If you have any concerns regarding our products, please contact ProductSafety@springernature.com

Printed and bound by CPI Group (UK) Ltd, Croydon, CR0 4YY

22/09/2026

02220958-0005